Mavis Sika Okyere
Gas Pipeline Integrity Management

Also of interest

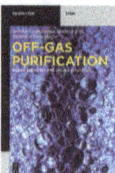

Mavis Sika Okyere

Gas Pipeline Integrity Management

Hazard Mitigation, In-Line Inspection, Risk Assessment,
Pipeline Operating and Maintenance

DE GRUYTER

Author

Dr. Mavis Sika Okyere
Pipelines and Stations Department
Ghana National Gas Company
225 Osibisa Road
Accra
Ghana
mavis.nyarko@ghanagas.com.gh

ISBN 978-3-11-162902-5
e-ISBN (PDF) 978-3-11-162974-2
e-ISBN (EPUB) 978-3-11-162989-6

Library of Congress Control Number: 2025933603

Bibliographic information published by the Deutsche Nationalbibliothek
The Deutsche Nationalbibliothek lists this publication in the Deutsche Nationalbibliografie;
detailed bibliographic data are available on the internet at http://dnb.dnb.de.

© 2025 Walter de Gruyter GmbH, Berlin/Boston, Genthiner Straße 13, 10785 Berlin
Cover image: Image taken during a pipe-to-soil potential test on Ghana National
Gas Limited Company pipeline
Typesetting: Integra Software Services Pvt. Ltd.

www.degruyter.com
Questions about General Product Safety Regulation:
productsafety@degruyterbrill.com

———

"To you, dear reader, for making this journey worthwhile."

Preface

The integrity of pipelines is a critical aspect of ensuring the safe and efficient transportation of resources. This book aims to provide a comprehensive guide to pipeline integrity management, covering the latest methodologies, technologies, and best practices in the field.

Over the years, the industry has faced numerous challenges, from aging infrastructure to evolving regulatory requirements. This book addresses these challenges by offering practical solutions and insights drawn from extensive research and real-world applications. It is designed to be a valuable resource for engineers, managers, and professionals involved in pipeline operations and maintenance.

I would like to express my gratitude to all the contributors and reviewers who have shared their expertise and knowledge. Their input has been invaluable in shaping the content of this book.

It is my hope that this book will serve as a useful reference and inspire further advancements in the field of pipeline integrity.

Sincerely,
Mavis Sika Okyere

https://doi.org/10.1515/9783111629742-202

Acknowledgments

I thank God for blessing me with the health, insight, and well-being needed to write this book. I'm grateful to my friends and family for their support and encouragement while working toward my goals. I am grateful to my husband, Yaw Okyere, and my children, for their unwavering love and patience. Their constant support has been invaluable throughout this journey.

I also wish to extend my gratitude to the publishing team, especially Robert Esposito, and Ute Skambraks, whose insights and guidance have been instrumental in shaping this book. To my colleagues and mentors who have inspired and challenged me, thank you for your wisdom and encouragement.

Finally, I would like to acknowledge all the readers who have shared this journey with me. Your interest and enthusiasm make this endeavor truly rewarding.

Thank you all.

https://doi.org/10.1515/9783111629742-203

Author biography

Dr. (Mrs.) Mavis Sika Okyere is a distinguished engineer and academic. She is currently a PhD candidate in Materials Science and Engineering at the University of Ghana. She holds a Doctorate in Oil and Gas Management from Wayne Park University in the USA, an MSc in Gas Engineering and Management from the University of Salford in the United Kingdom, and a BSc in Civil Engineering from Kwame Nkrumah University of Science and Technology in Ghana. Currently, she serves as the Assistant Manager of Pipeline Integrity at Ghana National Gas Limited Company.

Mavis has extensive experience implementing and managing pipeline integrity programs for onshore high-pressure gas pipeline systems. She is recognized for her meticulous engineering practices, thorough documentation, proficient technical report writing, and strong communication skills.

In her personal life, Mavis enjoys cooking, listening to music, and spending quality time with her family.

https://doi.org/10.1515/9783111629742-204

Contents

List of tables

https://doi.org/10.1515/9783111629742-206

List of figures

https://doi.org/10.1515/9783111629742-207

Abbreviations

CP	Cathodic protection
CSA	Canadian Standards Association
DNV	Det Norske Veritas
ECDA	External corrosion direct assessment
ESD	Emergency shutdown
ESDV	Emergency shutdown valve
HSE	Health, Safety, and Environment
HIC	Hydrogen-induced corrosion
ID	Internal diameter
IMP	Integrity management program
IGE	Institute of Gas Engineers
MOC	Management of Change
MOP	Maximum operating pressure
MAOP	Maximum allowable operating pressure
NDT	Nondestructive testing
NDE	Nondestructive examination
NACE	National Association of Corrosion Engineers
OD	Outer diameter
PE	Polyethylene
KPI	Key performance indicator
PSV	Pressure safety valve
ROW	Right of way
SCC	Stress corrosion cracking
SCADA	Supervisory control and data acquisition

https://doi.org/10.1515/9783111629742-208

1 Introduction

The potentially dangerous characteristics of certain gases, along with the existence of public water pipelines, make it essential to operate and maintain these pipelines effectively. It is crucial to adhere to established practices and procedures during the planning, operation, and inspection phases to ensure pipeline integrity. This guidance applies to all offshore pipelines and serves as a foundation for making ethical decisions. Failures in this area can pose risks to human safety, economic stability, and environmental health. This document outlines the fundamental requirements for data management, integration, and risk assessment strategies that are vital for preventing and mitigating these dangers.

Moreover, this guide promotes clear communication and accountability among all levels of management, support personnel, contractors, and facilitators, thereby enhancing safety measures. The program equips pipeline operators with essential information to effectively allocate resources for prevention, investigation, and risk mitigation, ultimately improving safety and minimizing accidents (refer to Figure 1.1).

This publication details how pipeline operators can identify and address risks to lower the probability and impact of incidents. It encompasses both performance-based and prescriptive integrity management approaches. A performance-based integrity management program relies on extensive data and comprehensive risk analysis, allowing operators to fulfill program requirements regarding inspection schedules, tools, and mitigation strategies. The prescribed process for establishing an integrity management program includes inspection, prevention, detection, and mitigation. Before implementing a performance-based integrity program, operators must conduct adequate inspections to gather the necessary information on pipeline conditions as dictated by the code-based program. The level of assurance provided by a performance-based system should be equal to or exceed that of a prescriptive system. Detailed requirements for both prescriptive and performance-based integrity management programs are outlined in specific sections of this document.

1.1 Objective

One of the key responsibilities of operators in pipeline systems is to maintain the integrity of the pipeline infrastructure. Gas pipeline operators primarily aim to ensure the safe and dependable functioning of gas pipelines, facilitating uninterrupted gas supply while minimizing any adverse effects on workers, the community, or the environment. An integrity management program is designed to implement a comprehensive set of processes focused on safety, operations, maintenance, assessment, and evaluation, all aimed at enhancing the protection of pipeline assets. This book serves as a resource for individuals and teams responsible for the planning, execution, and

https://doi.org/10.1515/9783111629742-001

enhancement of a pipeline integrity management program. Typically, a pipeline integrity team includes managers, engineers, and operations staff who possess specialized knowledge in detection, prevention, and mitigation strategies.

Key actions to consider include:

- Developing a pipeline integrity management system
- Integrating and harmonizing integrity processes
- Defining roles and responsibilities
- Implementing systems for continuous improvement
- Formulating strategies for mitigation, monitoring, and control

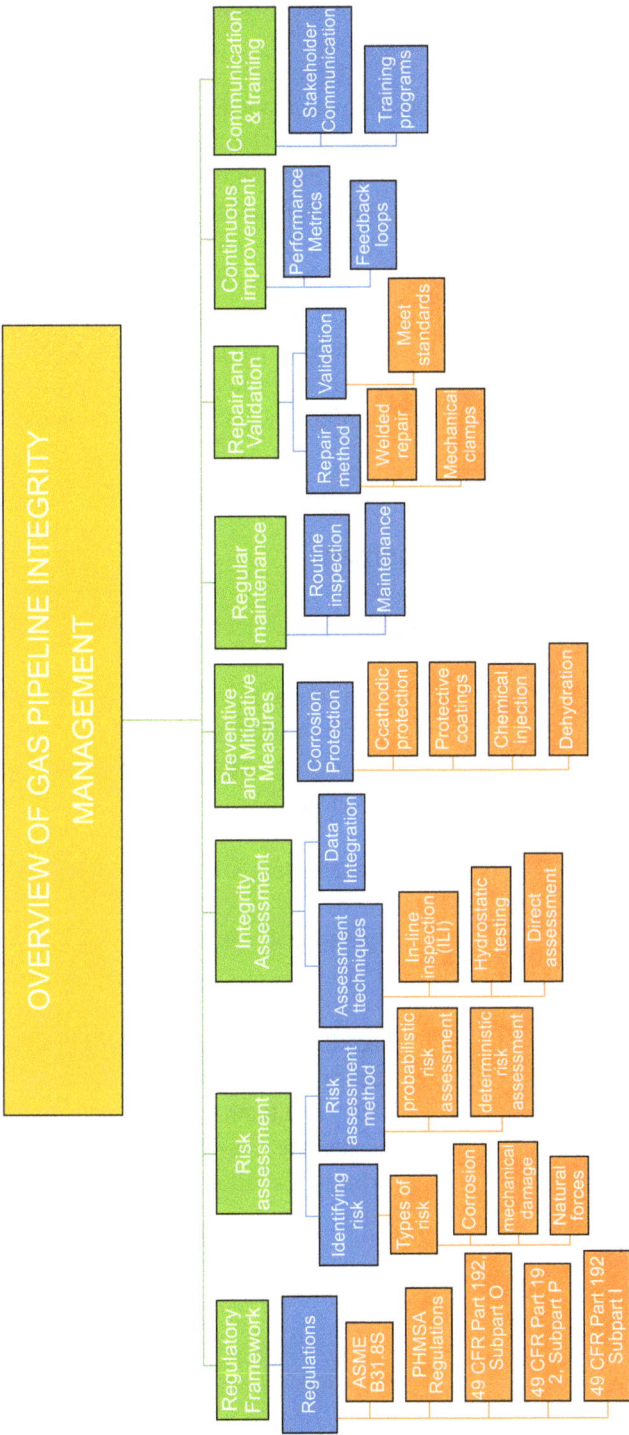

Figure 1.1: A brief overview of gas pipeline integrity management.

2 Hazard mitigation

A hazard refers to a characteristic or a combination of characteristics that may lead to potential loss. Examples of hazards include defects in materials, external damage, exposed pipelines, stress corrosion cracking, both internal and external corrosion, and the absence of necessary markings. The categorization of factors leading to hazards in pipeline systems is illustrated in Figure 2.1. This diagram outlines the various types of risks associated with pipelines, categorized into distinct groups based on their nature and origin. Each category highlights specific threats such as environmental conditions, mechanical failures, human errors, and external interferences, providing a comprehensive overview of the potential dangers that can compromise pipeline integrity and safety. Table 2.16 provides a summary of the available hazard mitigation methods.

2.1 Hazard identification and management

To identify and report hazards that may adversely impact pipeline systems, the following methodologies are utilized:
- Insights or results derived from integrity or operational monitoring including reports on right-of-way (ROW) evaluations and corrosion data
- Outcomes resulting from integrity assessments, such as failures identified during pressure testing or findings from in-line inspections (ILIs)
- Hazard identification worksheets
- Communication with landowners and other stakeholders, which may include reports related to planned ROW activities
- Reports generated from regularly scheduled operational meetings [1]
- Results obtained from formal risk analyses [1]

The data and methods employed for hazard identification consider the primary causes and subcauses specified in GS ISO 17776:2016 and LI 2189. Upon recognizing a potential integrity threat, it is critical to monitor the conditions that may exacerbate the threat and to take action to either eliminate or mitigate these conditions. Alternatively, the risk assessment team will conduct a thorough analysis of the risk and determine the requisite actions to be implemented.

https://doi.org/10.1515/9783111629742-002

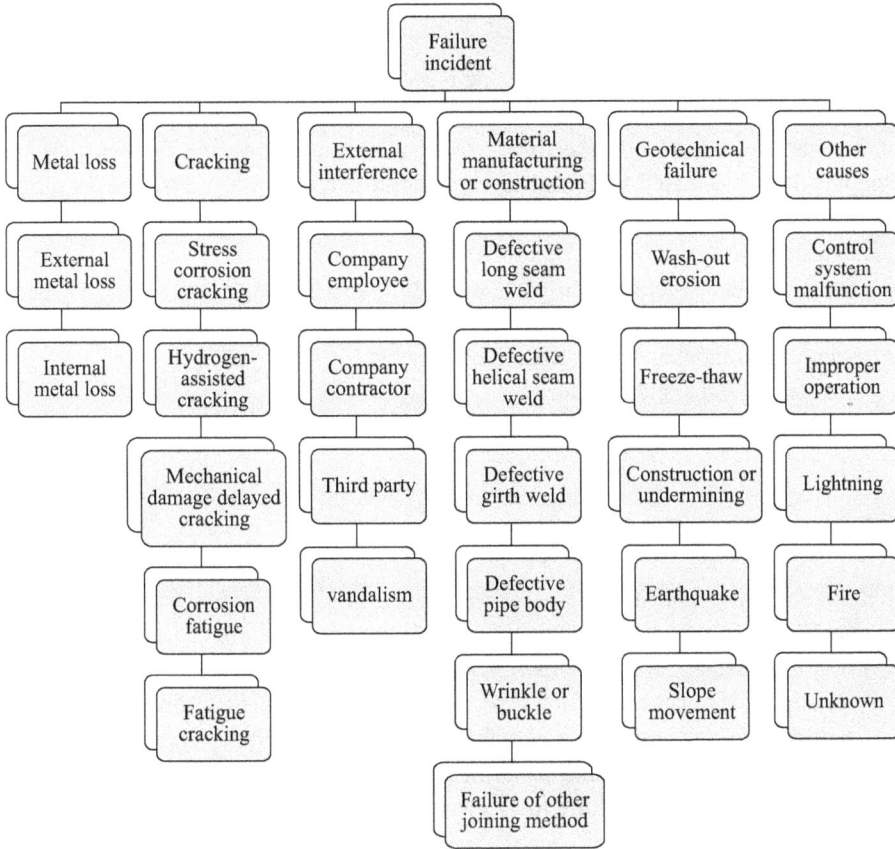

Figure 2.1: Classification of causes for pipeline hazards [2].

2.2 Mitigation of consequences of improper design or selection of materials

If an examination or risk assessment identifies a significant vulnerability attributable to inadequate design or materials, one or more of the following hazard mitigation strategies shall be enacted [1]:

– Conducting thorough inspections of in-line piping to evaluate the extent and nature of structural and material deficiencies
– Executing pressure tests on pipelines to detect defects in welds or materials

2.3 Addressing manufacturing deficiencies

Deficiencies in design and manufacturing may originate from a variety of sources including inadequacies in the steel production process, noncompliance of material properties with established specifications, errors occurring during construction, or complications associated with coating or cathodic protection (CP) systems. To effectively mitigate these risks, the implementation of the following measures is recommended [1, 2]:
- Establishment of comprehensive manufacturing specifications
- Definition of precise design specifications
- Adoption of independent verification processes
- Conducting independent audits
- Execution of pressure hydrostatic tests

2.4 Mitigation of external corrosion

In the event that a failure, direct examination, or risk assessment reveals a significant vulnerability due to inadequate CP or an excessive rate of external corrosion, it is imperative to implement one or more of the following risk mitigation strategies [1]:
- Conduct an external corrosion direct assessment (ECDA) study on pipelines suspected of having compromised external coatings in order to evaluate the effectiveness of CP along the entirety of the pipeline.
- Perform ILIs of the pipeline to ascertain the nature and extent of any external corrosion damage.
- Install additional test stations to facilitate the measurement of CP effectiveness in areas where instantaneous "OFF" readings are unattainable.
- Employ AC interruption technology to enhance the evaluation of polarized CP levels.
- Upgrade or install new CP systems to improve the distribution of protective current within the pipeline.

2.4.1 Common factors leading to external corrosion

Pipe coatings face numerous challenges that can undermine their protective capabilities. When subjected to high temperatures, these coatings may melt, soften, crack, or become brittle. Additionally, soil pressure from factors like the weight of backfill, thermal expansion-induced shear stress, pipe settling, or soil movement can cause the coating to separate or develop puckering. Excessive levels of CP in hot, humid soil conditions can also lead to blistering, particularly in fusion-bonded epoxy coatings [2].

Furthermore, localized increases in pH and the release of hydrogen during the application process can contribute to the degradation of the coating. Prolonged exposure to ultraviolet (UV) light before burial may cause coatings to crack or lose their integrity, especially if the coated pipe is left outdoors for an extended time. It is crucial to note that exposing fusible link epoxy coatings to UV radiation can lead to chalking, which is why consulting the manufacturer before application is essential. For aboveground pipe casings, protecting them from environmental elements is vital to avoid UV damage [8].

Table 2.1: Contributing factors and mitigation of external corrosion [3–5].

Contributor	Cause/source	Mitigation
Extreme operating temperature	– Failure of the coating – Disbondment of the coating	– Maintain the operating temperature below the maximum threshold established for the coating system [3, 4] – Select a coating system that possesses a temperature rating exceeding the expected operating temperature
Pipe displacement and soil stress	– Excessively high operating temperatures – Variability in operating temperatures – Insufficient support	– Appropriate design of the pipeline – Selection of coatings that fulfill the design specifications
Ground displacement and soil stress.	– Inconsistent soil conditions – Cycles of freezing and thawing	– Choosing a route – Stabilizing the soil – Selecting a coating
Improper handling and backfill	Rock deterioration	– Correct building methods – Choice of coatings
Poor joint coating	– Inadequate selection of joint coatings or mismatched pipe and joint coatings. – Incorrect application of joint coatings resulting from insufficient training, supervision, or inspection	– Appropriate design and engineering application guidelines or requirements – Skilled staff – Quality assurance in construction – Inspection of coatings

Table 2.1 (continued)

Contributor	Cause/source	Mitigation
Improper insulation	– Pipelines lack a corrosion barrier that separates the pipe from the insulation – Poor quality of joint coating leading to water infiltration	– Make certain that the coating system features a barrier resistant to corrosion – Adhere to the specified coating standards to uphold high quality in joint coating activities – Utilize injected molded foam at the joint in place of half-shells – Hire certified coating inspectors to evaluate and confirm the quality of the finished work
Concrete ballast and counterweight blocks	– Pipes in the concrete area that do not have adequate coating could be impacted – Coatings that are compromised	– The coating must be developed with the anchor's specifications considered – Coat the pipes prior to the concrete being poured – Verify the coating condition before positioning the anchor
Externally weight-coated pipe, and rock shielding	– These do not serve as corrosion barriers	Directly apply a corrosion barrier without holidays onto the pipe
Encased crossings or sleeved crossings	– The casing is in contact with the adjacent pipe. – The coating has been compromised.	– Install using centralizers made of nonmetallic materials – Ensure that the coating is entirely devoid of holidays – Avoid allowing water to infiltrate the casing
Trenchless crossings – no casing	– The coating was harmed during the installation process	– Install a corrosion barrier that is free during the holidays – Implement cathodic protection measures – Apply a coating that resists abrasion
Soil-to-air interface (risers)	– Impaired coating – Absence of coating	– Appropriate choice of coating – Position the coating cap over the interface – Conduct inspection and upkeep – Provide mechanical protection

Table 2.1 (continued)

Contributor	Cause/source	Mitigation
Insufficient levels of cathodic protection	– The cathodic protection system is operating in accordance with the standards set forth by NACE SP0169	Conduct an evaluation of the cathodic protection system and implement necessary adjustments
Electrochemical interference or cathodic disruption	– Electrochemical protection system or corrosion prevention system sourced internationally – Alternating current power transmission infrastructure	– Design an electrochemical protection system or corrosion prevention system effectively – Conduct thorough surveys and perform regular maintenance
overapplication of "cathodic protection" or "excessive electrochemical protection"	The system operated in an ineffective manner	Conduct a comprehensive assessment of the CP installed and implement necessary adjustments

2.4.2 Strategies for mitigating external corrosion during design and construction of pipelines

Table 2.2 presents various strategies for combating exterior corrosion that should be considered throughout the plan and installation phases of the pipeline design life [1, 2, 8].

Table 2.2: Approaches for reducing external corrosion during pipeline planning and installation [1, 3, 4].

Element	Practice
Coating – application process conducted at the manufacturing facility	– When selecting a surface treatment system, ensure that its design temperature is greater than the maximum operating temperature you anticipate for the application. This will help prevent potential failures or degradation of the coating over time due to thermal stress. – In the process of choosing the appropriate coating, consider the specific characteristics of the soil type in the environment where the coating will be applied. Different types of soil – such as water-saturated environments, sandy substrate, clayey ground, or rocky terrains – can significantly affect the adhesion, durability, and overall performance of the coating. Understanding these factors will aid in selecting a coating that provides optimal protection and longevity.

Table 2.2 (continued)

Element	Practice
Coating – plant applied thermally insulated pipe	– Think about a coating system that features a barrier against corrosion situated between the pipe and the insulation. – Consider water detection wires. – Safeguard the outer jacket coating system during installation in rocky soil conditions.
Coating – field applied at joints	– Choose a joint coating system that considers both present and anticipated operating conditions. – Choose a joint coating system that is compatible with the anticorrosion coating system of the pipe body. – Choose a joint coating system suitable for the field construction setting. – Apply proper surface preparation as advised by the coating manufacturer. – Create standards or specifications for coating application.
Joint type	If connections besides butt welds (e.g., zap-lok) are utilized, consider how it will affect cathodic protection.
Cathodic protection	Set up the cathodic protection system.
Inspection capability	– Set up or enable functionality for launching and receiving inspection tools. – Maintain uniform line diameter and wall thickness. – Implement piggable valves, flanges, and fittings.

2.4.2.1 External corrosion mitigation – operation

In cases where a failure, direct examination, or threat assessment reveals a significant vulnerability due to inadequate CP or unacceptable external corrosion levels [1], it is essential to implement one or more of the following risk mitigation strategies [1, 4]:

– Conduct an ECDA on pipes suspected of having compromised external coatings to evaluate the effectiveness of CP along the pipe's length.
– Conduct pigging of the pipeline to determine the nature and severity of damage resulting from pipeline outside corrosion.
– Set up extra monitoring stations to evaluate the efficiency of CP in areas where direct "OFF" potential measurements are impractical.
– Utilize alternating current interruption techniques for a more precise evaluation of polarized CP levels:
– Upgrade or install CP systems to enhance the distribution of current throughout the pipeline.

- Perform maintenance and repairs on the CP system to guarantee that the pipeline is sufficiently protected and complies with international standards such as ISO-15589:2015 or NACE SP0169-2013.
- Quickly address sections with worn or flaking paint by utilizing suitable materials after conducting a thorough inspection (like a DCVG survey).

Table 2.3 outlines various strategies for addressing external service corrosion throughout the lifecycle of the pipeline.

Table 2.3: Practices for mitigating external corrosion during operation [3, 4].

Element	Practice
Corrosion assessment	Understanding pipeline system coatings, evaluating operating temperature, assessing cathodic shielding potential, and reassessing CP system operation after line failure or system addition is crucial.
CP system maintenance	– Conduct an annual examination to confirm that there is enough CP current. – Inspect all insulating kits and joints; look for interference; and frequently check rectifiers and note outputs. – The CP system must be designed, installed, run, and maintained in compliance with NACE SP0169. – Make sure that rectifiers are regularly inspected to make sure that the desired current output is being achieved. – Cut down on needless rectifier downtime brought on by maintenance tasks. – Make sure CP surveys are carried out. – Address any disruptions, continuity bonding issues, isolation shortcomings, and other concerns quickly to ensure the CP system functions as designed; replace exhausted ground beds at the earliest opportunity; and enhance the CP system if more current is necessary to achieve the proper protection levels.
Inspection program	– Develop an inspection strategy. – Utilize root cause analysis results to modify corrosion mitigation and inspection programs.
Repair and rehabilitation	– Inspect to determine extent and severity of damage prior to carrying out repair or rehabilitation. – Based on inspection results, use CSA Z662 Clause 10.9.2 to determine the extent and type of repair required.
Failure analysis	– Recovery of an undisturbed sample of the damaged pipeline. – Conduct a thorough failure analysis. – Use the results of failure analysis to reassess CP system. – Measure pipe to soil potential at failure site.
Leak detection	Integrate a leak detection strategy.
Management of change (MOC)	– Implement an effective MOC process. – Maintain pipeline operation and maintenance records.

2.4.2.2 Corrosion detection methods

Corrosion monitoring techniques are essential for assessing the integrity of materials and structures subjected to corrosive environments. These methodologies allow for the systematic evaluation of corrosion rates and the effectiveness of corrosion prevention strategies. Implementation of reliable monitoring practices can significantly enhance maintenance planning and risk management, ultimately extending the lifespan of assets and ensuring operational safety.

Table 2.4 describes common methods to consider for detecting external corrosion and pipe coating degradation.

Table 2.4: Corrosion monitoring techniques [3, 6–8].

Technique	Description	Comments
Production tracking	Continuous observation of fluid temperature	High temperatures could harm the coating
Electrochemical protection	Maintenance, management, and operation of the electrochemical or cathodic protection	Consult NACE SP0169

2.4.2.3 Techniques for inspecting external pipeline corrosion

This section outlines various approaches and methodologies that can be utilized to effectively examine and identify signs of external corrosion on structures and materials. These techniques ensure a thorough understanding of the corrosion's extent and potential impact, allowing for timely maintenance and interventions.

Table 2.5 outlines various methods that are recommended for identifying external corrosion and the deterioration of coatings on pipelines.

Table 2.5: Inspection methods [4, 5].

Choices	Method
Cathodic protection effectiveness survey	Closed interval potential survey Annual system survey
Coating survey integrity	C – Scan coating conductance survey ACVG (pin-to-pin) and DCVGcoating survey
Leak detection	Ongoing surveillance of moisture levels

Table 2.5 (continued)

Choices	Method
Pigging	Magnetic flux leakage (MFL), ultrasonic testing, and eddy current testing are three widely used methods for inspecting materials and detecting flaws. Among these techniques, magnetic flux leakage is the most prevalent due to its effectiveness in identifying corrosion and other defects in pipelines and storage tanks. MFL works by detecting changes in the magnetic field around a conductive material, allowing for the pinpointing of irregularities. Ultrasonic testing utilizes high-frequency sound waves to penetrate materials and reveal internal defects, providing detailed information about the material's integrity. On the other hand, eddy current testing employs electromagnetic induction to identify surface and near-surface flaws, making it particularly useful for conducting quick assessments on conductive materials. Each of these techniques has its specific applications and benefits, contributing to a comprehensive approach to nondestructive testing.
Excavation and integrity assessments	Assessment, examination, and recording of the state of the coating and the pipe through physical exposure.

2.5 Internal corrosion mitigation

Based on Table 2.10, suppose assessments, including direct examination and threat evaluation, indicate a significant vulnerability to unacceptable internal corrosion costs. In that case, it is essential to implement one or more of the following risk mitigation strategies [1, 9]:
- Conduct ILIs of the pipeline to assess the extent and nature of internal corrosion.
- Review the current choice of pigs, their overall effectiveness, and the associated chemical treatment program.
- Explore alternative internal corrosion monitoring tools to enhance the detection of corrosion activity.
- Evaluate the feasibility of modifying operational conditions to lower the risk of corrosion.

2.5.1 Factors influencing corrosion occurring internally in gas pipeline systems

Several chemical components can lead to internal corrosion including
- Water (H_2O) content
- Hydrogen sulfide (H_2S) content
- Carbon dioxide (CO_2) content
- Dissolved solids

- Organic and inorganic acids
- Elemental sulfur and sulfur compounds
- Bacterial activity and their by-products
- Hydrocarbons
- The pH level of the fluid

In addition, various physical factors related to fluid dynamics contribute to internal corrosion such as
- Temperature
- Pressure
- Flow velocity
- Vibration
- Presence of entrained solids and liquids
- Deposits
- Flow characteristics and patterns (such as slug flow)
- The interaction of these factors along with other physical influences

Furthermore, the physical characteristics of the infrastructure, including pipes, vessels, pumps, compressors, and valves, can also affect internal corrosion. These factors include
- Material composition
- Residual or operational voltages
- Design features
- Presence of crevices
- Accumulations of deposits
- Surface conditions

Tables 2.6 and 2.7 detail the factors contributing to internal corrosion in gas pipelines, categorized by mechanisms and operational practices.

Table 2.6: Contributing factors – mechanism [1, 10].

Contributor	Cause/source	Mitigation
Hydrogen sulfide (H_2S)	– Produced with gas from the reservoir – Can be generated by sulfate-reducing bacteria – Iron sulfide scales tend to dominate when CO_2 to H_2S ratio is less than 20:1 (this limit is supplied as guidance only)	– Gas sweetening (amine treating) – Cleaning pigging – Injection of corrosion inhibitor – Dehydration – Small amounts of H_2S (i.e., in ppm level) can be beneficial as a protective FeS film can be established

Table 2.6 (continued)

Contributor	Cause/source	Mitigation
Carbon dioxide (CO_2)	– Produced with gas from the reservoir – Can be introduced as a frac medium	– Cleaning pigging – Dehydration – Injection of corrosion inhibitor
Bacteria	– Contaminated drilling and completion fluids – Contaminated production equipment – Produced fluids from the reservoir	– Cleaning pigging – Injection of biocides and inhibitors – Moisture control
Oxygen	– Ingress from compressors or vapor recovery units (VRU) – Introduced through endless tubing (ETU) well clean-outs – Ingress during line repairs or inspection – Injection of methanol – Frac fluids saturated with O_2	– Use gas blanketing and oxygen scavengers – Minimize oxygen ingress and/or inhibit the pipeline – Optimize methanol injection and/or use inhibited methanol – Frac fluid to be deaerated or O_2 scavenged
Water holdup	– Low gas velocity or poor pigging practices allow water to stagnate in the pipelines – Low gas pressure (e.g., <1,000 kPa) may not have the gas density to push water even at higher flow rates – Absence of water separation equipment leads to water wet pipelines – Deadlegs or inactive service	– Install pigging facilities and maintain an effective pigging program – Remove water at the wellsite by separation or dehydration – Control corrosion through effective inhibition – Remove inactive deadlegs – Effectively pig lines as soon as the wells become inactive
Polysulfides	– May be produced with formation water from sour reservoirs – Polysulfides are water soluble molecules – Not detected in standard water analysis	– Install pigging facilities and maintain an effective pigging program – Implement a corrosion inhibition program
Chlorides	– Produced with formation water – Can be the result of spent acid returns from well stimulation	– Dehydration – Injection of corrosion inhibitor

Table 2.6 (continued)

Contributor	Cause/source	Mitigation
Solids deposition	– Includes sands, wax, asphaltenes, and scales – Loose iron sulfide accumulations are commonly formed in sour systems – Can originate from drilling fluids, workover fluids, and scaling waters – May include corrosion products from down hole or upstream equipment – Insufficient gas velocities or poor pigging practices	– Install pigging facilities and maintain an effective pigging program – Initially, use well site separators to tank and truck liquids to minimize the effects of work over and completion activities on the pipeline – Scale inhibition – Install gas conditioning systems
Methanol	– Excessive quantities of injected methanol (methanol injection should be limited to a 1:1 water/methanol ratio or the amount required for hydrate inhibition, methanol can contain up to 70 mg/L dissolved O_2) – Use of uninhibited methanol	– Avoid overinjection of methanol – Effective pigging and inhibition – Remove free water – The addition of gas dehydration or line – Heaters can reduce or eliminate the need for methanol usage – Use inhibited methanol
Elemental sulfur	– Produced from a reservoir or formed in the system – Formed due to the reaction of H_2S and oxygen – Without oxygen, thermodynamic instability/solubility due to pressure and temperature change could also generate – Elemental sulfur. – Sulfur deposition is more prevalent in liquid hydrocarbon-free systems	– Install pigging facilities and maintain an effective pigging program – Implement a corrosion inhibition program – Implement sulfur solvent treatments – Eliminate oxygen ingress

Table 2.7: Contributing factors – operating practices [1, 3, 10].

Contributor	Cause/source	Mitigation
Drilling and completion fluid	– Introduction of spent acids and kill fluids – Introduction of solids – Introduction of bacteria – Introduction of O_2 and CO_2	– Produce well-to-well site separator, tankage, and trucking water until drill and complete fluids and solids are recovered – Supplemental pigging and inhibition of pipelines before and after work over activities

Table 2.7 (continued)

Contributor	Cause/source	Mitigation
Critical gas velocity	Critical gas velocity is reached when there is insufficient flow to sweep the pipeline of water and solids	– Design pipeline to exceed critical velocity – Establish operating targets based on critical gas velocity to trigger appropriate mitigation requirements (pigging, batch inhibition, etc.)
Detrimental operating practices [10]	– Ineffective pigging – Ineffective inhibition – Intermittent operation – Inadequate pipeline suspension practices – Commingling of incompatible produced fluids – Flow back of work-over fluids into the pipeline – Deadlegs due to changes in production or operation of pipelines	– Design pipelines to allow for effective shut-in and isolation – Develop and implement proper suspension procedures, including pigging and inhibition – Establish acceptable operating parameters – Test for fluid incompatibilities
Management of change (MOC)	– Change in production characteristics or operating practices – Well re-completions and workovers – Lack of system operating history and practices – Changing personnel and system ownership	– Implement an effective MOC process – Maintain integrity of pipeline operation and – Maintenance history and records – Reassess corrosivity on a periodic basis

2.5.2 Suggested strategies for reducing internal corrosion

Table 2.8 outlines the suggested strategies for minimizing internal corrosion during the design and construction phases of pipelines. Meanwhile, Table 2.9 details the recommended approaches for addressing internal corrosion during the operational phase of gas pipelines [3, 10, 11].

2.5.3 Procedures for mitigating internal corrosion during operations

Internal corrosion pertains to the deterioration occurring within pipelines. This form of corrosion is frequently triggered by the presence of various substances such as carbon dioxide (CO_2), hydrogen sulfide (H_2S), water, natural acids, microorganisms, and other reactive compounds. These substances typically engage with the pipe's inner surface through anodic and cathodic processes. The outcomes of these reactions may deposit within the

Table 2.8: Suggested strategies – design and construction [1, 3, 10].

Element	Recommended practice
Materials of construction	– Use normalized EW line pipe that meets standard requirements of standards such as CSA Z245.1 steel pipe – Consider using corrosion-resistant nonmetallic materials such as HDPE or composite materials – Use sour service steel pipe for sour gas pipelines as per the requirements of codes such as CSA Z662
Dehydration	– Install gas dehydration facilities – Ensure dehydration units are operating properly
Gas conditioning	– Install inlet separator – Install gas treating (sweetening)
Water removal	– Install water separation and removal
Pipeline isolation	– Install valves that allow for effective isolation of pipeline segments from the rest of the system – Install the valves as close as possible to the tie-in point – Install blinds for effective isolation of inactive segments
Deadlegs Pipeline sizing	– Design and construct a system to avoid or mitigate the effect of deadlegs – Establish an inspection program for existing deadlegs – Design a pipeline system to maintain flow above critical velocity – For pipelines that operate below the critical velocity ensure corrosion mitigation programs are effective for the conditions
Pigging capability	– Install or provide provisions for pig launching and receiving capabilities – Use consistent line diameter and wall thickness – Use piggable valves, flanges, and fittings
Inspection capability	– Install or provide the capability for inspection tool launching and receiving – Use consistent line diameter and wall thickness. – Use piggable valves, flanges, and fittings

pipeline, potentially forming a protective layer that inhibits further corrosion. Conversely, in some instances, these products do not precipitate, resulting in accelerated corrosion rates. The extent of internal corrosion is influenced by factors such as the concentration of corrosive substances, temperature, flow rate, and the material of the pipeline [2, 12].

Within the pipeline, a phenomenon known as "black powder" can occur. This is because the gas, which is expected to be dry, may still contain some water vapor that can condense and lead to corrosion [3]. The resultant corrosion products can accumulate as a powder within the pipeline, potentially obstructing or damaging valves and metering devices. Black powder in gas pipelines typically comprises corrosion by-products, but it may also include residues from mill scale, welding spatter, and other debris. To minimize the formation of black powder, it is essential to maintain low moisture levels and implement strategies to reduce oxygen contamination. This in-

Table 2.9: Recommended practices – operation [10].

Element	Recommended practice
Completion and workover practice	Produce wells to surface test facilities until drilling and completion fluids and solids are recovered
Corrosion assessment	– Evaluate operating conditions (temperature, pressure, water quality) and prepare a corrosion mitigation program – Communicate corrosion assessment, operating parameters, and the mitigation program to all key stakeholders, including field operations and maintenance personnel – Reassess corrosivity on a periodic basis and subsequent to a line failure
Corrosion mitigation and monitoring	– Develop and communicate the corrosion mitigation and monitoring program to all key stakeholders including field operations and maintenance personnel – Note: Ensure personnel understand their responsibilities and are accountable for implementation and maintenance of corrosion management programs – Develop pipeline suspension and discontinuation procedures
Failure analysis	– Recover an undisturbed sample of the damaged pipeline – Conduct a thorough failure analysis – Use the results of failure analysis to reassess corrosion mitigation program
Repair and rehabilitation	– Inspect to determine extent and severity of damage prior to carrying out repair or rehabilitation – Based on inspection results, use CSA Clause 10 to determine extent and type of repair required – Implement or make modifications to corrosion control program after repairs and failure investigations so that other pipelines with similar conditions are inspected and mitigation programs revised as required
Leak detection	Integrate a leak detection strategy
Management of change (MOC)	– Implement an effective MOC process – Maintain pipeline operation and maintenance records

cludes avoiding the use of air for drying after hydrotesting and eliminating potential sources of oxygen ingress during operations [2, 12].

The oil and gas sector employ various strategies to protect against internal corrosion including

– internal coatings
– gas quality management such as dehydration
– chemical treatments including corrosion inhibitors and biocides
– equipment upkeep such as cleaning pigs
– buffering techniques
– internal CP (applicable only for tanks)

To reduce internal corrosion in gas pipelines, it is crucial to limit contaminants in the gas entering the system. Implementing stringent quality control measures is vital. Regular analysis of product samples should be conducted, and any solids or liquids extracted from the pipelines must be tested for corrosive agents like sulfur compounds and bacteria. The protective strategies against internal corrosion will be elaborated upon in subsequent sections. One effective approach to mitigate internal corrosion is to monitor and control the quality of gas entering the pipeline. Additionally, by routinely sampling and analyzing the gas, liquids, and solids extracted from the pipeline for corrosive contaminants and corrosion products, operators can identify potential corrosion issues, understand their causes, and formulate appropriate anticorrosion strategies [2, 12].

In the event of a failure, if direct examination or risk assessment indicates a significant risk of unacceptable internal corrosion rates, the following risk mitigation measures should be considered:

- Conducting ILIs to evaluate the nature and extent of internal corrosion
- Reviewing the selection and effectiveness of existing pigs and chemical programs
- Exploring alternative internal corrosion monitoring technologies to better gauge corrosion activity [1]
- Conducting feasibility studies to modify operating conditions to lower the risk of corrosion [1]

2.5.3.1 Product monitoring

Determine the most important variables:

- pH,
- temperature,
- pressure,
- flow,
- water chemistry,
- bacteria,
- suspended particles,
- chlorine,
- oxygen,
- dissolved metals, and
- chemical residues.

Product monitoring and internal corrosion mitigation strategies should be taken into consideration to maintain a low risk of internal corrosion.

It is crucial to keep an eye on the temperature of the liquid (gas). The coating may be harmed by high temperatures. Controlling the quality of the gas entering the pipeline is one method to lessen the likelihood of internal corrosion. The pipeline operator can also ascertain whether hazardous corrosion may occur by routinely sampling and examining the gas, liquids, and solids extracted from the pipeline for the presence and concentration of any corrosive contaminants, including bacteria, as well as the presence of corrosion products. Determine the reasons for corrosion and create strategies to prevent it [3, 12].

2.5.3.2 Internal coatings

The main objective of internal pipe coating is to reduce pipe friction and internal corrosion. These properties result in lower operating and installation costs, higher product purity, and higher throughput. A protective coating for pipe cores has long been recognized as bringing countless savings to pipeline operators. The inner coating provides protection against corrosion, abrasion; reduce the cost of scrubbers, strainers, "piggy banks," and other types of pipe cleaning services. It ensures the purity of the product, prevents contamination by corrosion products, significantly reduces maintenance and labor costs, provides protection of the inside of the pipe from the accumulation of deposits (limestone or paraffin), and as has been well-proven, substantially increases the "passability" of the product and can be achieved with pipes with an internal coating. For both liquid and gas pipelines, the cost of internal pipeline coating can in most cases be justified only based on reduced operating costs [3, 7].

Early field tests with gas pipelines showed that, depending on the pipe and flow characteristics, it was possible to increase throughput by 5–10% with a 24″ pipe [3, 11]. A possible 1% increase may justify the cost of interior painting; this measured increase appears to represent an economic incentive [3]. However, for most applications, contract delivery and/or manufacturing considerations preclude the use of an internal coating based on either increased product permeability or reduced piping costs. Pipe coating is therefore of relatively little value as a means of increasing capacity. Achievable internal pipe surface roughness varies depending on the liner material used, but for maximum hydraulic improvement, wall roughness for gas pipes should be around 5–10 μm [9]. Known advantages associated with the internal coating of liquid piping include reduced maintenance and lower wax build-up [13]. Up to 25% reduction in wax formation is achievable by applying an internal coating [9, 18]. For liquid systems, the economic benefits are greater for smaller pipe diameters, while for gas systems, the benefits are greater for larger pipe diameters. In most cases, the benefits are greater for gaseous than for liquid systems [13].

2.5.3.2.1 Epoxy pipe coating

Two-component epoxy interior pipeline coatings are used in the interior of pipelines used in the transportation of dehydrated natural gas, wet gas, crude oil, sour crude oil, salt water, potable water, fresh water, petroleum products, and many chemicals. Such specialist epoxy coatings for inner tubes have now been available for several years and, through practical experience, can be applied in adequate film thickness with the required resistance characteristics. The two main methods for applying interior coatings are

a. Spraying
b. In situ coating

2.5.3.3 Chemical injection
In the normal oil and gas industry, the following chemical additives are stored and added to the process to ensure reliable and safe operation.

2.5.3.3.1 Corrosion inhibitor
These chemicals prevent corrosion of pipelines, pipes, or tanks. The corrosion inhibitor works by creating a passivation layer on the metal that prevents the corrosive substance from reaching the metal.

A corrosion inhibitor reduces the rate of corrosion of metal exposed to this environment. Inhibition is used internally on carbon steel pipes and vessels as an economical corrosion control alternative to stainless steels and alloys, coatings or nonmetallic composites and can often be implemented without process interruption. Corrosion inhibitor or other corrosion protection chemicals such as monoethylene glycol (MEG) injected into the system. The main industries using corrosion inhibitors are oil and gas exploration and production, oil refining, chemical manufacturing, heavy manufacturing, water treatment, and additive manufacturing [3].

2.5.3.3.2 Scale inhibitor
Scale is a deposit of an insoluble inorganic mineral. Common oilfield scales include calcium carbonate and barium sulfate. Scale deposits in process units such as tubes and heat exchangers impede or block fluid flow. A scale inhibitor inhibits the formation and deposition of scale [14].

Scale-inhibiting chemicals that are applied up or down the wellhead and are generally divided into four categories:
- Miscible with oil
- Completely water-free
- Emulsified
- Solid

Depending on the mineral content present in the water, the duration of the work and the operational needs, the chemicals can be applied continuously or during scale compaction.

2.5.3.3.3 Biocides
These prevent microbiological activity in oil production systems. Uncontrolled activities of bacteria, algae, and fungi are problematic in oil field operations. For example, bacterial activity such as sulfate-reducing bacteria leads to the production of H_2S, which leads to tank acidification, metal corrosion, health hazards, and filter clogging. Typical uses include diesel tanks, produced water (after hydrocyclones), sludge, and ballast tanks. Biocides are used to inhibit and eliminate microbiologically influenced corrosion caused by the corrosive action of microbes. The biocide is injected into the

pipeline in a stream of nonelectrolytic carrier. In many cases, the biocide is added to the buffering agent so that only one addition to the gas stream is required. Other active agents may also be added such as film-forming agents that help form a passive barrier on the pipe surface and agents that promote electrolyte evaporation. Many of these agents are expensive and, depending on gas flow at the time of injection or use, may or may not reach the site where electrolytes and microbes are trapped [3] The biocides discussed here are used in many industries such as the oil and gas industry. Microbial control in the oil and gas industry is primarily used to prevent the harmful effects of microbial growth on production equipment, pipelines, and tanks.

2.5.3.3.4 Antifoam
It is a chemical additive mixed with industrial process fluids to prevent foaming in industrial process fluids. Some of the commonly used defoamers or defoamers are insoluble oils, glycols, polydimethylsiloxanes, silicones, octyl alcohol, aluminum stearate, and sulfonated hydrocarbons are used as antifoam/defoamers [15].

Defoamers prevent the formation of foam during oil processing by reducing the surface tension of the liquid. It is particularly used in units such as separators where foaming prevents effective separation of gas from liquids. Antifoam is used to remove foam that can cause overflow, clogging, corrosion, or electrical shorting. If foam is produced in any industrial process liquid, it can cause serious problems such as defects in surface coatings due to the formation of air bubbles in the surface coatings. Because of this, unevenness occurs and a smooth surface coating is not achieved.

2.5.3.4 Dehydration
Dehydration is the most prevalent method of protecting against internal gas corrosion in pipes (also in pipelines for liquids containing oil and free water or other electrolytes) [3, 16–19]. Dehydration removes condensation and free water, which would otherwise allow internal corrosion to occur when water droplets precipitate from the gas stream and form liquid pools at the pipe's bottom or stick to the pipe's top [3, 16–19]. Topside corrosion is uncommon in dry gas streams. Complete dehydration is quite successful, but because the systems are not 100% efficient or 100% trustworthy, there is always the chance of introducing water and other electrolytes into the pipeline [3, 16–19].

Gas dehydration is a process that removes water from the gas. The water content of a gas stream is calculated as the weight of water per volume of gas (mg/Sm^3) [3, 16–20]:
–   The dew point at reference pressure is the temperature at which water vapor condenses in a gas
–   Water concentration in gas (ppm)

The following determines the ability of a gas to hold water:
–   Higher temperatures
–   Higher temperatures increase the gas's ability to contain more water [20]

- Lower pressure
- Low pressure enhances water retention
- High pressures lead to the presence of CO_2 and H_2S

Gas with these impurities should hold more water at higher pressures and a correction should be made when calculating the water content of such a gas stream, especially when the gas mixture contains more than 5% H_2S and/or CO_2 at pressures above 4,800 kPa [3, 16].

Typical dehydration strategies for both upstream and downstream sections of the gas value chain are
- Dehydration of glycol in a gas processing plant (see Figure 2.2)
- Partial dehydration at the wellhead and subsequent additional steps to meet contract specifications
- Chemical injection at the wellhead with subsequent dehydration at the central discharge point [3, 16]
- Complete dehydration at each wellhead

2.5.3.4.1 Reason for gas dehydration
Dehydration is utilized to manage the dew point of water. This is crucial for achieving the specifications for product water dew point and preventing operational issues. It may lead to the condensation of water:
- Corrosion: Carbon steel piping is susceptible to corrosion from CO_2 and H_2S
- Hydrate prevention

2.5.3.4.1.1 Common methods of gas dehydration
The primary technologies employed for gas dehydration include
a. glycol dehydration through physical absorption (using MEG, diethylene glycol (DEG), triethylene glycol (TEG), or tetra-ethylene glycol (TREG));
b. adsorption on a fixed bed utilizing molecular sieves;
c. low-temperature separators paired with a glycol injection system (TEG).

2.5.3.4.1.2 Glycol dehydration
Glycol dehydration is a method used for drying liquids by extracting water from natural gas and natural gas liquids. It is the most prevalent and cost-effective way to eliminate water from these streams. The glycols typically used in the industry include TEG, DEG, MEG, and TREG [3]. TEG is the most often used glycol in industrial applications [3]. The procedure entails absorbing water vapor using a glycol solution within a contact tower. Highly concentrated glycol solutions, such as TEG, are used to physically extract water from the gas, and the glycol solution is then regenerated for future use. Figure 2.2 depicts a standard dehydration unit, though architecture may vary from one plant to another [3].

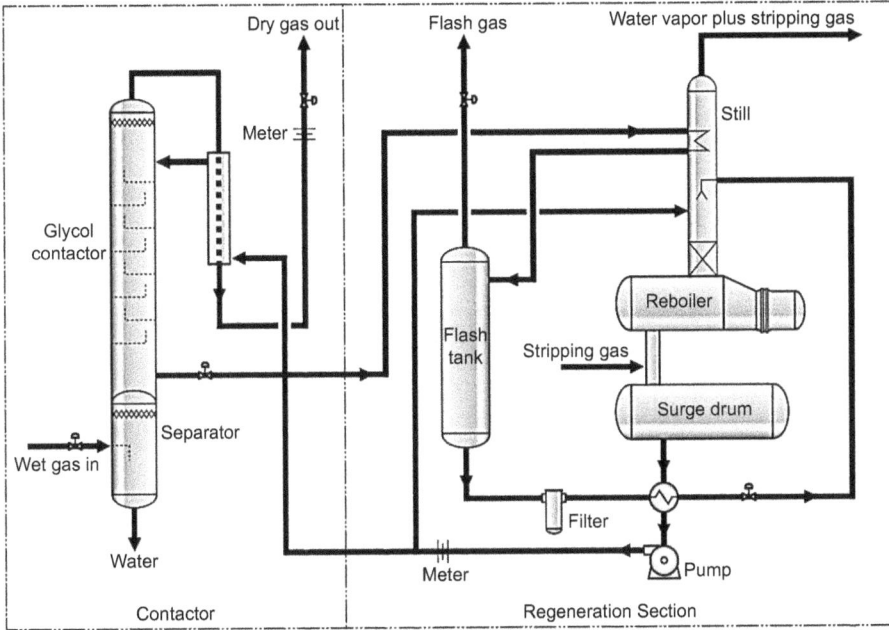

Figure 2.2: Example glycol dehydration unit (GDU) [3].

2.5.3.4.1.3 Adsorption on solid bed (e.g., molecular sieves)

Adsorption separation techniques utilize a solid bed material that has a high specific surface area. Various solid desiccants possess the physical attributes needed to extract moisture from natural gas (refer Figure 2.3).

Commercial solid desiccants can be classified into three main types:

- Gels (such as alumina or silica). Silica gel is commonly employed to dehydrate and recover i-C5 + from a natural gas stream. When silica gel is used for dehydration, the water dew point can reach around −60 °C. Silica gel is the simplest desiccant to regenerate, and while it does absorb heavier hydrocarbons, these can be removed more easily during the regeneration process.
- Alumina (activated aluminum oxide) serves as a solid desiccant that can achieve a water dew point as low as 73 °C.
- Molecular sieves (composed of aluminosilicates) offer the greatest water absorption capacity, with the ability to meet a water dew point of 100 °C. As selective absorbers, they do not absorb heavier hydrocarbons, but they demand the most energy for desiccant regeneration.

At least two vessels are required for the solid bed adsorption operation, with one removing moisture while the other regenerates and cools. In general, an adsorption cycle might last anywhere from 8 to 24 h [3].

In conclusion, there are three main commercial ways for regenerating saturated desiccant materials [3]:
- Temperature swing adsorption (TSA)
- Pressure swing adsorption (PSA)
- Inert purge stripping

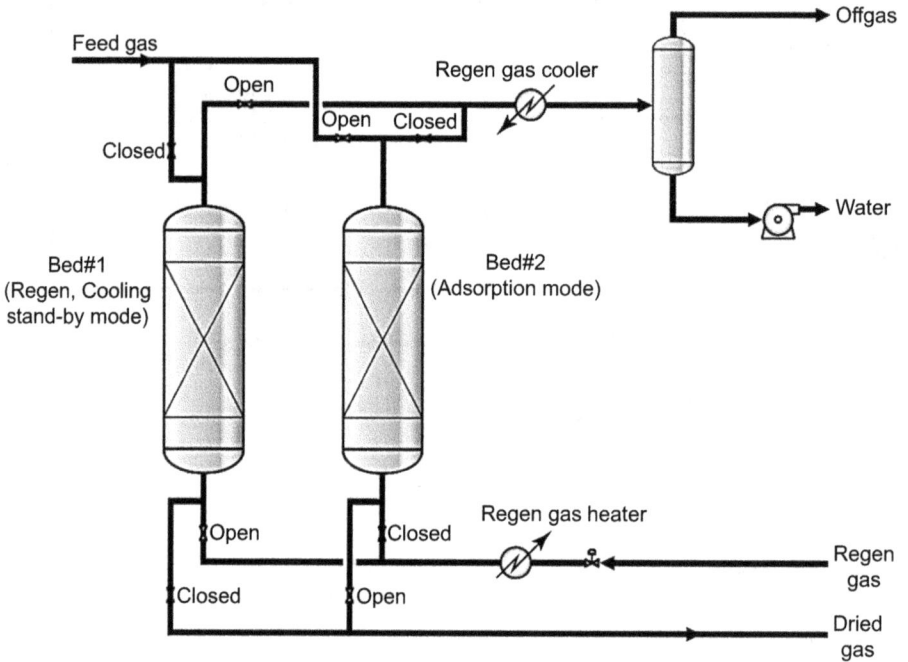

Figure 2.3: Typical gas dehydration by adsorption process [3].

The evaluation of the solid desiccant system needed can be determined by several factors:
1. The required dew point indicates that molecular sieves deliver the lowest dew point necessary before cryogenic processing.
2. The dimensions of the drying beds should account for the limited space available on the coastal platform. It is essential to examine the absorption capacity of each desiccant under the dryer's operating conditions.
3. The composition of the gas is important as certain desiccants absorb larger hydrocarbon molecules. These larger molecules should be released during the regeneration process and become part of the heating gas stream. Consequently, these larger, potentially valuable products will not be recoverable in processing facilities.
4. The duration of the cycles influences the drying time before regeneration, impacting the size of the dryers. A shorter drying time allows for smaller dryer sizes,

but quicker cycles increase the likelihood of mechanical failures in switching systems. The ideal drying duration is roughly 24–48 hours.

5. For higher regeneration temperature needs, a larger regeneration heating device will be necessary. Again, it is crucial to factor in space limitations.

2.5.3.4.1.4 Low-temperature separator (LTS) with glycol injection system

A low-temperature separator equipped with a glycol injection system is utilized when the dew point for water or hydrocarbons cannot be achieved, and it also helps to prevent the formation of hydrates. This system is commonly employed when the installation of a complete gas dehydration unit or solids adsorption unit is not financially feasible. The gas is cooled to allow for the condensation of both water and hydrocarbons (NGLs). Cooling is accomplished through methods such as mechanical cooling, the Joule-Thomson (J-T) effect, or a turboexpander. The dew point for both water and hydrocarbons is influenced by the operating temperature of the cold separator. Glycol injection acts to inhibit hydrate formation, which could obstruct the equipment. This process is depicted in Figure 2.4 [3].

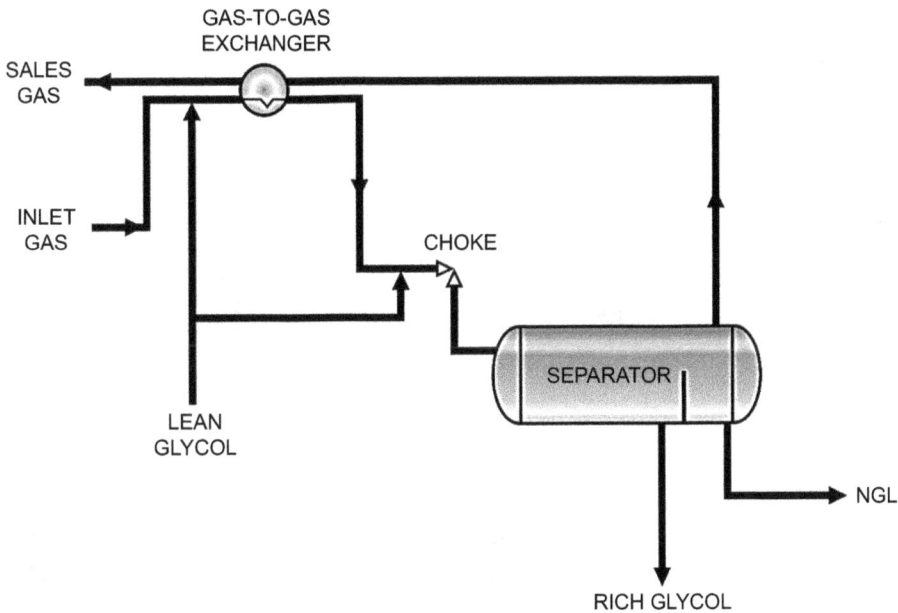

Figure 2.4: Gas dehydration module [3].

2.5.3.5 Cleaning pigs

The term "pig" refers to a pipeline inspection gauge. Pigging is a technique for In-Line Inspection (ILI) that involves inserting devices known as "pigs" into the pipeline for cleaning and/or inspection purposes. A pig is a tool inserted into a pipeline to carry out various maintenance tasks (see Figure 2.5). These tasks encompass cleaning the pipe and inspecting it, which provides insights into the pipe's condition and identifies the extent and location of any issues, such as corrosion and pipe sealing through plugging. The pig is dispatched from the pig launching unit and is propelled by the product flowing through the pipeline. On the opposite end, it is caught by a pig receiving station [3].

The primary advantages of pipe pigging include the removal of dirt from the pipeline and the gathering of data regarding its condition. This data allows users to identify issues that can aid repair personnel in maintenance tasks.

Cleaning pigs are capable of effectively channeling liquids and corrosive solids into pig traps, enabling their removal from the pipes. Regular maintenance should ensure that any fluid accumulations are directed away from low areas and, if performed correctly, eliminated from the entire pipeline. Additionally, cleaning pigs should transport solids and extract them from the pipe via a pig trap situated at the end.

Figure 2.5: Pipeline pigs [3].

These pigs are specifically designed to eliminate deposits that could hinder material flow:

1. Mechanical cleaning is a widely recognized and accepted method among pipeline operators for extracting deposits from pipelines. The pig is sent through the pipe multiple times to dislodge sediment until minimal sediment is detected at the receiving station. However, confirming that the pipe is entirely clean can be challenging.
2. Advanced chemical cleaning [3], known as "Advanced Dry Cleaning," is quickly becoming the standard in the industry. The combination of chemical cleaning with mechanical pigs effectively removes larger quantities of dirt in fewer passes. Chemical cleaning involves using liquid cleaning agents mixed with a diluent (such as water, diesel, methanol, and isopropyl alcohol) to create a cleaning solution that can be pushed through the pipes by pigs. For instance, to eliminate hy-

drocarbon hydrates from a pipeline system, a batch of either methanol or glycol is passed through the pipeline with the assistance of a pig. The packing size should be adequate to lower the hydrate formation temperatures to below the maximum expected deposition temperature, allowing the liquid to adhere to the pipe wall rather than being pushed ahead by the pig. Hydrate removal from raw natural gas pipelines, which may lack gas scrubbers, is typically achieved by increasing the feed rate of the hydrate point depressant. Shock dosing should be administered briefly at a rate approximately five times higher than the usual injection rate. The selection of the appropriate pipeline cleaners should be based on the following criteria [19, 39]:

- A neutral pH.
- The ability to permeate and penetrate deposits.
- The original design specifications of the cleaner and its historical performance.
- Health, safety, and environment (HSE) considerations.

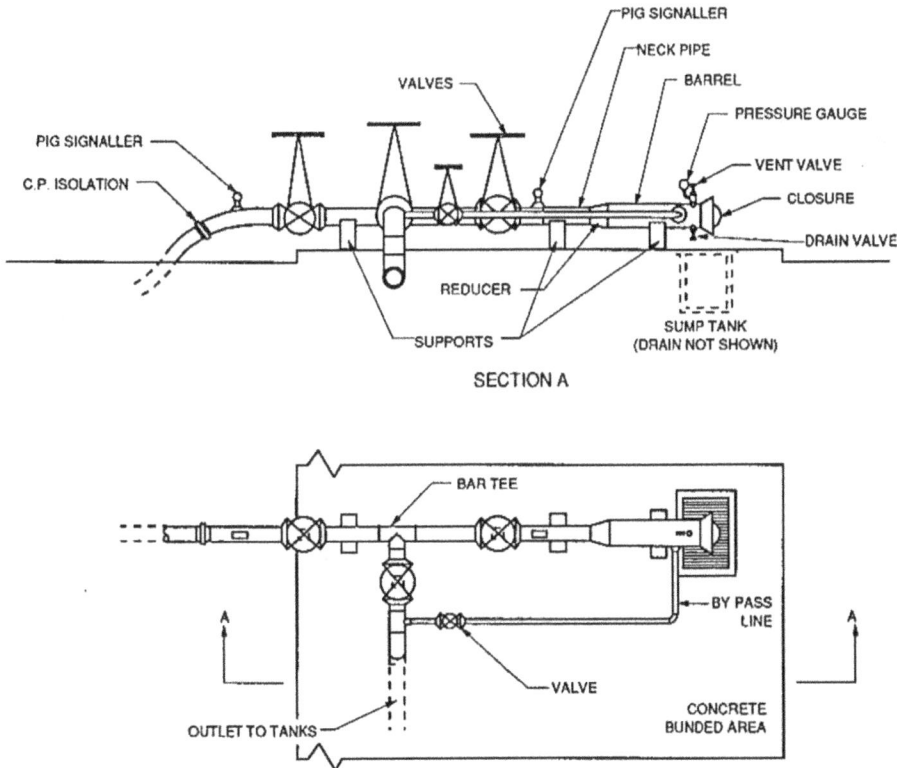

Figure 2.6: Typical pig receiving facilities [3].

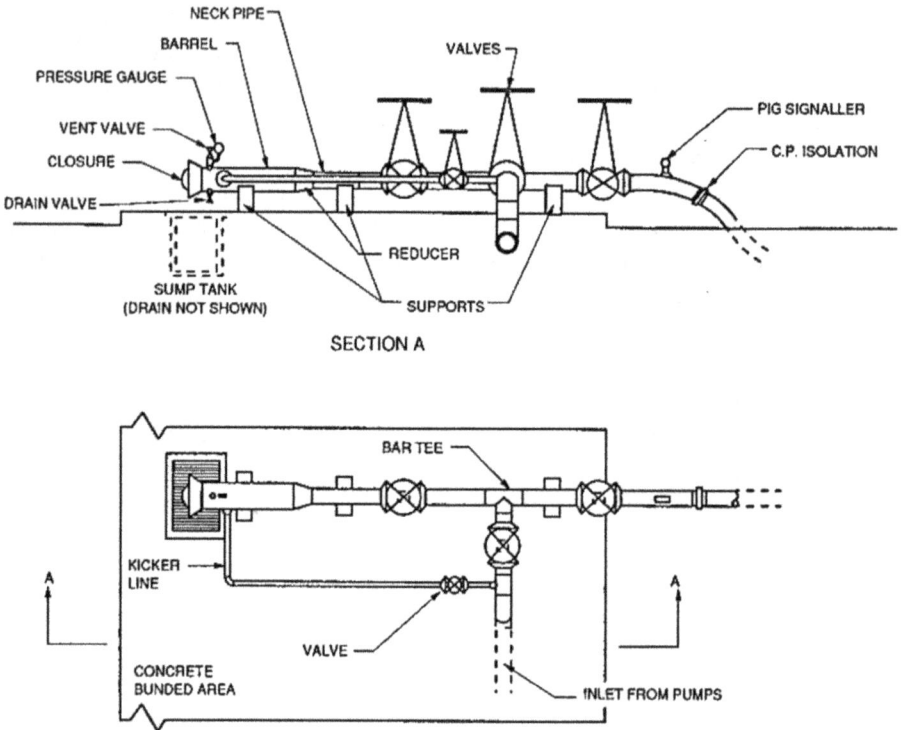

Figure 2.7: Typical pig launching facilities [3].

2.5.3.6 Buffering

A buffering agent is a weak acid or base utilized to keep the acidity (pH) of a solution close to a specific value even after additional acids or bases are introduced. This indicates that the role of a buffering agent is to stabilize the pH level and prevent it from fluctuating rapidly when acids or bases are introduced to the solution [3, 17].

Buffering agents alter the chemical composition of liquids that remain in pipes, making them effective in avoiding internal corrosion [3, 16–19]. Furthermore, a buffering agent, such as a weak or dilute alkaline solution, can significantly reduce the corrosiveness of any stagnant liquid, principally by raising its pH above seven (neutral), converting it from acid to basic [3, 16–18]. Alkaline solutions usually do not affect steel. However, damping is rarely very efficient because it is difficult to cover the full surface of the pipe [3, 16–19].

The application of buffers or other chemicals that adjust pH can also change the environment, potentially stopping the proliferation of bacteria that lead to microbial corrosion [3, 16–19].

Table 2.10: Practices for mitigating internal corrosion during operation [3, 4, 10, 21].

Technique	Description	Comments
Production monitoring	Monitor fluid (gas) temperature	The temperature of both the soil and the pipeline can create favorable conditions for the deterioration of pipeline materials. Although liquid and gas pipelines exhibit slightly different operating temperature characteristics, both remain vulnerable. For instance, gas pipelines can experience temperatures ranging from a peak of 40 °C immediately after leaving the compressor station to as low as 5 °C at greater distances from the station.
Cleaning pigging	Periodic pigging of pipeline segments to remove liquids, solids, and debris	– Pigging is among the most effective methods for controlling internal corrosion. – It serves as a valuable technique for both cleaning pipelines and minimizing the risk of bacteria colonization and underdeposit corrosion. – Proper selection of pig type and size is crucial. – Facilities for the launching and receiving of pigs are essential.
Batch corrosion inhibition (chemical treating)	– Regular application of a batch corrosion inhibitor is essential for creating a protective barrier on the interior surface of the pipeline – The initial batch treatment is crucial during the commissioning of the pipeline, following new construction, repairs, or temporary suspensions – Batching should be conducted after any activities that may disrupt the protective films, such as inspections, line repairs, or workovers	– Acts as a barrier between corrosive elements and the surface of the pipe. – The application procedure is crucial in determining effectiveness; factors such as the volume and type of chemical, the diluent used, contact time, and application interval all play a significant role. – Effectiveness may be diminished in the presence of existing pitting, particularly if the pits are deep and narrow. – The inhibitor can be effectively applied between two pigs to clean the pipe and deposit the inhibitor. – It should be used alongside pigging to remove liquids and solids, as the inhibitor is most effective when applied to a cleaned pipe.

Table 2.10 (continued)

Technique	Description	Comments
		– Some corrosion inhibitors may have reduced efficacy in the presence of top-of-the-line corrosion. – It is important to note that the inhibitor must be applied to a clean pipe to achieve optimal results when used in conjunction with pigging.
	It is important to note that large diameter lines may necessitate specialized design considerations and procedures to ensure the integrity of the batch slug. Batch programs encompass a variety of variables, including personnel, chemicals, and applications, and require careful management to guarantee effective implementation and thorough performance monitoring.	
Continuous corrosion inhibition (chemical treating)	The continuous injection of a corrosion inhibitor is a strategic approach utilized to mitigate the corrosivity of fluids during transport. This process not only helps in reducing the degradation of materials caused by corrosive agents but also aids in the formation of a protective barrier film. This barrier serves as a shield, enhancing the longevity and integrity of the transportation system by preventing direct contact between the fluids and the surfaces of the infrastructure. Employing this method is essential for maintaining operational efficiency and minimizing maintenance costs associated with corrosion-related damage.	– The effectiveness of a corrosion inhibitor may be diminished when it is unable to adequately reach and cover the entire inner surface of a pipe, particularly in systems that have accumulated dirt and contaminants. In such instances, using a batch application method can prove to be more effective, as it allows for a more concentrated treatment to contact the surfaces directly. – The design of the corrosion prevention program is crucial for achieving optimal results. Key factors in this design include the selection of appropriate products, establishing relevant performance criteria, and understanding the specific production characteristics of the system in which the inhibitors are being utilized. These considerations help tailor the approach to meet the unique needs of the system.

Table 2.10 (continued)

Technique	Description	Comments
		– Implementing a well-structured program can significantly enhance the mitigation of top-of-the-line corrosion (TLC). By addressing the unique challenges of the pipeline environment and ensuring proper application of the inhibitors, it's possible to prolong the lifespan of the infrastructure and reduce maintenance costs.
Biocide chemical treating	Regular application of a biocide is essential for the eradication of bacteria within the pipeline system.	– Aid in managing bacterial growth or eradicating bacteria in systems identified to harbor them. – Employ in conjunction with pigging (for line cleaning) to improve effectiveness. – Batch application is generally the most effective method (e.g., down-hole application results in continuous treatment of produced fluids flowing into the pipeline). – The use of inadequately chosen biocides may result in foam formation, which can pose significant operational challenges.
Oxygen control	– Implement gas blanketing and utilize oxygen (O_2) scavengers. – Refrain from purging test equipment into the pipeline. – Optimize the injection of methanol and/or consider employing inhibited methanol. – Conduct batch treatments for pipelines following repairs, inspections, and hydrotesting.	The introduction of oxygen into an environment can significantly increase the corrosion potential of materials, particularly metals. This heightened corrosion potential arises because oxygen acts as an electron acceptor in electrochemical reactions, which can lead to the accelerated deterioration of metal surfaces. Additionally, the presence of oxygen can contribute to the formation of sulfur compounds when it interacts with sulfur-containing substances in the environment.

Table 2.10 (continued)

Technique	Description	Comments
		This reaction not only further exacerbates corrosion processes but can also lead to the generation of harmful by-products that may have detrimental effects on both materials and surrounding ecosystems. Understanding these interactions is crucial for developing effective strategies to mitigate corrosion and protect infrastructure

2.5.3.7 Corrosion monitoring techniques

Table 2.11 describes the most common corrosion and service monitoring techniques conditions associated with internal corrosion in gas pipelines.

Table 2.11: Internal corrosion monitoring technique [3, 10, 21].

Technique	Description	Comments
Gas and oil analysis	Continuous monitoring of gas composition is essential, particularly for detecting the presence of hydrogen sulfide (H_2S) and carbon dioxide (CO_2). In cases where either of these compounds is identified, it becomes important to conduct a thorough analysis of the properties of liquid hydrocarbons, as this information can provide valuable insights into the quality and safety of the hydrocarbons being processed. This analysis may involve examining factors such as viscosity, density, and the overall chemical composition of the liquid to assess its behavior and compatibility within the system.	– It is essential to fully comprehend the content of acid gases present in the reservoir, and this understanding should be regularly updated through periodic assessments to ensure accurate data. These assessments will help determine any changes in acid gas levels over time, which can significantly impact operational and safety protocols. – Additionally, the trend of reservoir souring should be closely monitored. This involves tracking changes in the composition and quality of the reservoir fluids, particularly with respect to the increasing presence of souring agents such as hydrogen sulfide (H_2S). Ongoing monitoring will allow for timely interventions and adjustments to extraction techniques, ultimately safeguarding the integrity of the reservoir and minimizing environmental impacts.

Table 2.11 (continued)

Technique	Description	Comments
Water analysis	Continuous oversight of water quality is essential, focusing on key parameters such as chloride levels, the presence of dissolved metals, concentrations of suspended solids, and residual chemicals from various processes. This ongoing monitoring involves regular sampling and analysis to ensure that water meets safety and regulatory standards. Sensors and testing methods will be employed to detect any fluctuations in these elements, allowing for timely interventions to mitigate potential environmental or health impacts.	– Alterations in water chemistry are likely to affect corrosion potential. – Trends in the concentration of dissolved metals, such as iron and manganese, can serve as indicators of changes in corrosion activity. However, monitoring the iron-manganese ratio may be less effective in H_2S systems. – Chemical residuals can be utilized to verify application levels and fluctuations in water production. – The selection of sampling locations and adherence to proper procedures are essential for obtaining accurate results.
Production monitoring	Continuous monitoring of production conditions is crucial for maintaining optimal performance and safety in industrial operations. This involves regularly measuring and analyzing key parameters such as pressure, temperature, and flow rates. By keeping track of pressure levels, we can ensure that they remain within safe operational thresholds to prevent equipment failure or dangerous incidents. Temperature monitoring helps in maintaining the integrity of processes and products, while flow rate assessments provide insights into the efficiency of production systems. Together, these factors contribute to a comprehensive understanding of the production environment, allowing for timely adjustments and interventions when necessary.	– Variations in operating conditions, such as temperature, pressure, and fluid composition, must be taken into account as they can significantly affect the corrosion potential of materials used in systems. Monitoring these changes is crucial for maintaining the integrity of the equipment and preventing premature failure. – Detailed production data is essential for evaluating the susceptibility of materials to corrosion. This assessment can be made by analyzing factors such as fluid velocity and corrosivity, which are directly related to the chemical nature of the fluid and its flow characteristics. By understanding these parameters, more effective corrosion mitigation strategies can be implemented.

Table 2.11 (continued)

Technique	Description	Comments
Mitigation program compliance	It is essential to conduct continuous monitoring and assessment of the implementation, execution, and documentation processes related to the mitigation program. This involves regularly reviewing the effectiveness of the strategies being executed, ensuring that they align with the established objectives, and thoroughly documenting all actions taken and their outcomes for accountability and future reference.	– The reliability of chemical pumps, along with effective management of inhibitor inventory, is essential in contexts where a mitigation program necessitates ongoing chemical injection to prevent corrosion. Ensuring that these pumps operate without failure is crucial, as any downtime can compromise the integrity of the entire system. – To achieve optimal results, a corrosion mitigation program needs to be not only properly implemented but also continuously maintained. This includes regular monitoring and adjustments to chemical dosing, as well as thorough inspections of equipment to detect any early signs of corrosion or potential failures in the injection process. – It is important to conduct a comprehensive evaluation of the consequences resulting from any deviations or noncompliance with the established mitigation program. Assessing the impact on corrosion rates and potential damage is vital for informed decision-making and for implementing corrective actions to minimize risks associated with equipment failure or safety hazards.
Corrosion coupons	The evaluation process is employed to determine the susceptibility of materials to both general corrosion and localized pitting corrosion. This assessment is crucial for understanding how these materials will perform in specific environments. Additionally, it serves to analyze the effectiveness of various corrosion mitigation programs that have been implemented, allowing for	– Analyzing trends in coupon data can provide valuable insights into variations in corrosion activity, helping to identify potential issues before they escalate. By closely monitoring this data over time, it's possible to detect patterns that may signal an increase in corrosion rates or the effectiveness of corrosion control measures.

Table 2.11 (continued)

Technique	Description	Comments
	adjustments and improvements to be made based on the results. By systematically conducting these assessments, we can better ensure the longevity and integrity of structures exposed to corrosive conditions.	– It is essential to utilize coupons alongside a range of other monitoring and inspection techniques to obtain a comprehensive understanding of corrosion processes. Combining multiple approaches ensures a more robust assessment, allowing for a thorough evaluation of the corrosion environment and the performance of materials in service. – The effectiveness of coupon monitoring relies heavily on several key factors, including the type of coupon selected, its placement within the system, and how the resulting data is interpreted. Careful consideration of these elements is crucial to ensure accurate and meaningful results that contribute to informed decision-making regarding corrosion management strategies. Properly positioned and appropriately chosen coupons can enhance the validity of the corrosion assessments made, leading to better maintenance practices and extended asset life.
Bio-spools	The system is utilized to detect the presence of bacteria and assess the effectiveness of biocides in eliminating microbial contamination. It helps ensure that the biocides applied are working efficiently and maintaining a safe environment by regularly monitoring bacterial levels.	– The placement of bio-spools and the interpretation of data are essential for the effective application of these methods. – Bio-spools should be utilized alongside other monitoring and inspection techniques. – Solids extracted from pipelines, known as pig yields, can be analyzed for the levels of sessile bacteria. – The presence of bacteria is a critical factor to consider.

Table 2.11 (continued)

Technique	Description	Comments
Electrochemical monitoring	There are several comprehensive methods utilized for analyzing and measuring corrosion, each with its unique principles and applications. One such method is electrochemical noise measurement, which detects fluctuations in the electrical signals generated by a corroding surface, providing insights into the corrosion processes. Linear polarization is another approach, involving the application of a small voltage to a corroding electrode to measure the resulting current, which helps in determining the corrosion rate. Additionally, electrical resistance techniques involve assessing changes in the resistance of a material over time, allowing for monitoring of corrosion as it occurs. Hydrogen foils or probes are effective in measuring the concentration of hydrogen in the environment, which can be indicative of corrosion activity, particularly in certain metal types. Lastly, the field signature method analyzes the electromagnetic fields produced by corroding structures, offering a noninvasive means to evaluate the integrity of materials in situ. Each of these methods provides valuable data for understanding and managing corrosion effectively.	– The careful selection of devices, their appropriate placement, and the accurate interpretation of the collected data are essential components for the successful implementation of these monitoring methods. These factors directly influence the reliability and validity of the results obtained. – Various data collection methodologies may be employed, including continuous monitoring, where data is gathered in real time over an extended period, and intermittent collection, which involves taking measurements at specific intervals. Both approaches have their advantages and can be chosen based on the specific needs of the monitoring process. – Electrochemical monitoring techniques can be effectively utilized alongside other monitoring and inspection methods. Integrating these approaches enhances the overall effectiveness of the assessment by providing a more comprehensive understanding of the system being studied.

2.5.3.8 Corrosion inspection techniques

Table 2.12 outlines typical methods to evaluate internal corrosion in gas pipelines. Please note that since localized corrosion is the primary failure mode in sour gas pipelines, relying solely on hydrotesting may not adequately verify the integrity of the pipeline.

Table 2.12: Internal corrosion inspection technique [3, 10, 21].

Options	Technique	Comments
In-line inspection	– Various nondestructive testing (NDT) techniques are available for assessing the integrity of materials and structures. Among these, magnetic flux leakage (MFL) is the most widely utilized method. This technique excels in detecting surface and subsurface flaws by measuring the leakage of magnetic fields from a magnetized object, making it particularly effective for inspecting pipelines and storage tanks. – In addition to MFL, ultrasonic testing (UT) and eddy current testing are also prominent options in NDT. Ultrasonic tools use high-frequency sound waves to identify internal defects, providing precise measurements of thickness and detecting flaws that may not be visible externally. Meanwhile, eddy current testing employs electromagnetic induction to detect surface defects and measure conductivity in conductive materials, which is valuable for assessing material properties. – Each of these methods – MFL, ultrasonic, and eddy current – offers unique advantages and applications, enabling engineers and inspectors to choose the most suitable technique based on specific inspection needs and conditions.	– An effective approach for accurately assessing the location and severity of corrosion. – In-line inspection can effectively detect both internal and external corrosion defects. – The tools are available in either self-contained or tethered configurations. – The pipeline must be designed or modified to facilitate in-line inspection. – To conduct a tethered tool inspection, it is often necessary to excavate bell holes and cut into the pipeline.

Table 2.12 (continued)

Options	Technique	Comments
Nondestructive examination	Ultrasonic testing, radiographic inspection, and various other nondestructive testing (NDT) methods can be employed to quantify localized metal loss effectively.	– Prior to conducting nondestructive evaluation (NDE), it is essential to perform an assessment to identify potential corrosion sites. – Refer to NACE SP0110 for the Wet Gas Internal Corrosion Direct Assessment Methodology for Pipelines, or NACE SP0206 for the Internal Corrosion Direct Assessment methodology for pipelines carrying normally dry gas. – Utilizing multifilm radiography serves as an effective preliminary screening tool before proceeding with ultrasonic testing. – NDE is frequently employed to validate in-line inspection results, assess corrosion at excavation sites, and evaluate aboveground piping. – It is important to understand the practical limitations of NDE methods and the factors that can impact their accuracy.
Video camera/ boroscopes	A visual inspection tool designed to identify and locate internal corrosion within structures or systems. This tool enhances the detection process by providing clear, visual cues that allow inspectors to evaluate the condition of materials, ensuring early intervention and maintenance to prevent further deterioration.	– This method is commonly employed to identify and assess the presence of corrosion damage on various surfaces. However, it is important to note that it does not provide a clear assessment of the severity of the corrosion detected, which can complicate maintenance decisions. – Additionally, the effectiveness of this technique may be constrained by the length of the inspection distance, limiting its application to localized areas rather than extensive surfaces. – Furthermore, it lacks the capability to directly measure the depth of corrosion pits, which is crucial for evaluating the potential impact on structural integrity and safety.

Table 2.12 (continued)

Options	Technique	Comments
Destructive examination	A three-dimensional physical representation showcasing various segments extracted from the pipeline system. Each section highlights the structural components, materials used, and relevant connections, offering a comprehensive view of the pipeline's architecture and functionality.	When planning assessments for pipeline integrity, it is crucial to identify and prioritize locations that are particularly susceptible to specific failure modes. This proactive approach allows for a more focused inspection and maintenance strategy. Additionally, it is recommended to consult the NACE SP0110 Wet Gas Internal Corrosion Direct Assessment Methodology for Pipelines, which provides a comprehensive framework for evaluating corrosion risks in wet gas environments. For pipelines that primarily transport normally dry gas, reference should be made to the NACE SP0206 Internal Corrosion Direct Assessment Methodology. This methodology outlines effective assessment techniques and best practices aimed at identifying and mitigating internal corrosion in these types of pipelines.

2.6 Third-party damage mitigation

If, through comprehensive risk assessments or direct inspections, you determine that the pipeline could potentially be at risk of damage from third-party activities, it is essential to implement a series of mitigation strategies. These strategies should be put in place before and during the construction phase, and they should remain in effect throughout the entire operational lifespan of the pipeline. Consider the following detailed mitigation approaches:

1. Enhanced surveillance and monitoring: Establish robust surveillance systems to monitor activities around the pipeline. This can include aerial inspections, ground patrols, and the use of advanced technologies like drones or remote sensing to detect any unauthorized work or potential threats.
2. Buffer zones: Create a buffer zone around the pipeline that restricts development and construction activities. Mark these zones with signage to inform the public and contractors of the pipeline's presence and the associated risks.

3. Public awareness programs: Launch educational initiatives aimed at local communities and businesses, informing them about the pipeline's location, the importance of safeguarding it, and the potential hazards associated with interference.
4. Third-party engagement: Establish communication and collaboration with neighboring landowners and third parties operating in proximity to the pipeline. This can involve sharing information about the pipeline's specifications and the risks of encroachment, along with best practices for safe operations around it.
5. Regular maintenance and inspections: Schedule regular maintenance checks and inspections of the pipeline, focusing on areas identified as higher risk based on previous assessments. This proactive approach can help address vulnerabilities before they lead to significant issues.
6. Emergency response plans: Develop and implement thorough emergency response plans that outline procedures for addressing damage from third-party activities. This should include training for the response teams and regular drills to ensure preparedness.

By integrating these strategies, you can significantly enhance the safety and integrity of the pipeline and reduce the likelihood of damage caused by external factors during construction and throughout its operational life [3, 22]:

1. Public awareness [3]
2. Route selection: avoid populated areas [3]
3. Burial depth: increased depth of burial [3]
4. Provide additional protection [3]
5. Pipeline surveillance [3]
6. Improve signage [3]
7. ROW intrusion detection [3]
8. Notification system [3]
9. Safety management [3]
10. Improve material quality [3]
11. Decreased stress [3]
12. Design factor: use of lower design factor [3]

These mitigation measures are described below [3].

2.6.1 Public awareness

Prior to commencing pipeline construction, it is recommended to consult with local authorities, regulatory agencies, and community groups. This engagement should aim to secure the consent of at least 85% of landowners, advocacy organizations, and service providers along the pipeline route, indicating an approval in principle for the project. However, it is important to note that this approval in principle does not carry legal weight and does not confer any construction rights for the pipeline. Stakeholders should be approached early in the process and given the opportunity to provide feedback on the proposed project [3].

When considering the initiatives for public education outlined in Table 2.1 of API RP 1162, several effective strategies can be employed to enhance community awareness and safety regarding relevant issues:

- Developing informational handouts and brochures: Create a variety of educational materials that convey essential safety information. These handouts can include graphics and illustrations to make the information more accessible and engaging for the public.
- Participating in exhibitions and presentations: Actively engage in local community events, fairs, and expos to present valuable information. This can include setting up booths that provide interactive displays, demonstrations, and opportunities for attendees to ask questions and learn more about safety initiatives.
- Distributing environmental brochures and information: Provide targeted brochures that focus on environmental safety and responsibility. Ensure that these materials highlight the importance of protecting local ecosystems and detail specific actions the public can take to contribute positively.
- Coordinating with local authorities: Collaborate with municipal and regional officials to align educational efforts with community safety priorities. This can involve joint initiatives or partnerships that strengthen outreach and reinforce messages through credible local channels.
- Communicating with landowners and occupiers: Establish direct lines of communication with landowners and those who occupy land in the vicinity of safety initiatives. This can include personalized outreach, meetings, or informational sessions to discuss pertinent safety practices and regulations.
- Responding to general public inquiries: Set up a dedicated channel for the public to ask questions and receive timely, accurate responses regarding safety and environmental concerns. This could be through a hotline, email, or social media, ensuring accessibility for all community members.
- Promoting through advertisements: Utilize various advertising venues, such as local newspapers, radio stations, and online platforms, to disseminate safety messages broadly. Tailor advertisements to resonate with specific audience demographics for more effective outreach.
- Producing safety-related videos: Create informative and engaging video content that addresses safety topics relevant to the community. These videos can be shared on social media, community websites, or local television to reach a wider audience and reinforce educational messages visually.

Each of these initiatives plays a crucial role in building awareness and promoting safety in the community, fostering a well-informed public capable of making safer choices.

Suppose the pipeline owner has the authority to purchase land or obtain rights for pipeline placement without the consent of the landowner or occupant. In that case, this should be explored and utilized when necessary. Monthly public awareness initiatives can be organized to inform the community. Guidelines for developing and executing improved public awareness programs can be found in standards such as API RP 1162, 49 CFR 195.440, and 49 CFR 192.616.

2.6.2 Route selection

Effective pipeline routing is crucial for creating a pipeline system that optimizes material costs, construction efficiency, and safety. When planning new pipelines, it is important to consider the potential for increased population density in the future to prevent related issues. Therefore, it is advisable to route pipelines a sufficient distance away from densely populated areas, such as towns, cities, and villages, to mitigate any future impacts related to expansion [3].

When routing pipelines onshore, it is essential to consider a variety of general requirements and parameters to ensure project feasibility and compliance. These include:

- Legislative requirements: Compliance with local, state, and federal regulations governing pipeline construction and operation, including obtaining necessary permits and adhering to industry standards.
- Land ownership and compensation: Identifying land ownership along the proposed pipeline route and negotiating compensation agreements with landowners to secure the necessary rights for construction and operation.
- Environmental sensitivity: Assessing the potential environmental impacts of the pipeline, including effects on wildlife habitats, ecosystems, and water sources, and implementing measures to mitigate any adverse effects.
- Habitation proximity: Evaluating the distance of the pipeline from residential areas and populated regions to minimize risks to communities and comply with safety regulations.
- Land use and classification: Considering existing land use patterns, such as agricultural, commercial, or industrial activities, and ensuring that the pipeline routing aligns with local zoning laws and land classification.
- Topography: Analyzing the physical landscape through which the pipeline will be routed, including elevations, slopes, and geological conditions that may affect construction and long-term stability.
- Crossings of obstacles: Identifying natural and man-made obstacles, such as rivers, roads, and other utilities, that the pipeline will need to cross, and planning appropriate techniques for safe and effective crossings.

- Safety codes and requirements: Adhering to all applicable safety codes and regulations to ensure the safe construction and operation of the pipeline as well as minimizing risks to workers and the surrounding community.
- Construction access: Evaluating the access points for construction crews and equipment and ensuring that there are sufficient routes to safely transport materials and personnel to the pipeline location.
- Existing pipelines, utilities, and overhead power lines: Conduct a survey of existing infrastructure in the area to avoid conflicts with other pipelines, utility lines, and overhead power cables, which may require additional planning and coordination.

The parameters involved are extensive, and each pipeline must be evaluated on a case-by-case basis. Land use planning factors should be considered not only in the routing of new pipelines – aiming to minimize proximity to populated areas whenever possible – but also in making decisions regarding proposals for construction near existing pipelines [3, 22].

2.6.3 Burial depth: increased depth of burial

As shown in Figure 2.8, onshore pipelines are generally installed underground for several important reasons including
- safety,
- protection against potential damage,
- reduced environmental and visual impacts, and
- minimizing the area of land "sterilized" by the pipeline, which refers to restrictions on land usage for construction and other activities.

The depth at which a pipeline is buried is influenced by various factors, as outlined in the design codes, with details provided in Table 2.13. In urban areas, such as towns and cities, and near active zones like major roadways, historical data suggest that increasing the soil cover over the pipeline to a maximum of 2 m can significantly mitigate risks from third parties who typically operate within a depth of 1–1.5 m. An evaluation of damage records from various pipeline transmission systems reveals that the risk of damage decreases by a factor of 10 when the cover is raised from 1.2 to 2.2 m [3, 17].

However, while the advantages of increased cover are clear, there are also higher costs associated with the additional excavation and the increased volume of earth that must be moved due to the deeper burial of the pipeline. For example, for a 36-in pipeline, excavation might reach depths of up to 3.5 m if a cover of 2.2 m is required in certain locations. This situation presents further hazards for construction workers, including the risk of trench collapses and difficulties related to water accumulation within the trench [3, 17].

Figure 2.8: Natural gas pipeline being buried [3, 23].

Table 2.13: Minimum requirements for the coverage of buried pipelines [3, 8, 24, 25].

Location	IGE/ TD/1 edition 1	IGE/TD/ 1 editions 2–4	IGEM/ TD/1 edition 5	PD 8010–1:2015	ASME B31.8	AS 2885.1	L.I. 2189	
							Normal soil	Consolidated rock
All	0.9 m (3 ft)	1.1 m						
Rural areas	–	–	1.1 m	0.9 m	0.61 m (Class 1) 0.76 m (Class 2)	0.75 m	762 mm (Class 1)	457 mm (Class 1)
Suburban areas	–	–	1.1 m	1.2 m	0.76 m (Classes 3 and 4)	0.9 m	914 mm (Classes 2–4)	610.9 mm (Classes 2–4)
Roads	–	–	1.2 m	1.2 m	0.91 m	–		
Water courses, canals, rivers	–	–	1.2 m	1.2 m		1.2 m	914 mm	610 mm
Railways	–	–	1.4 m	1.4–1.8 m	0.91 m	–		

Table 2.13 (continued)

Location	IGE/ TD/1 edition 1	IGE/TD/ 1 editions 2–4	IGEM/ TD/1 edition 5	PD 8010–1:2015	ASME B31.8	AS 2885.1	L.I. 2189 Normal soil	Consolidated rock
Rocky ground	–	–	–	0.5 m	–	0.9 m (W) 0.6 m (T1, T2) 0.45 (R1, R2)		

2.6.4 Additional protection

To enhance the protection of the pipeline and mitigate risks associated with third-party activities, it is recommended to implement a range of protective measures. As illustrated in Figure 2.9, these may include increasing the thickness of the pipeline wall, as well as installing concrete slabs, tiles, high-density polyethylene plates (such as overpipe high-density polyethylene plates), high-tensile netting, or protective sleeves in high-risk areas.

The installation of reinforced concrete slabs, durable tiles, or strong steel plates directly above the pipeline serves a critical purpose. These barriers act as a deterrent against accidental damage caused by excavation or construction activities. In practice, if excavation work occurs, workers will encounter the slab, tile, or plate before reaching the pipeline itself. This proactive approach ensures that any potential damage is contained to the protective barrier, significantly reducing the likelihood of harm to the pipeline and maintaining its integrity. By anticipating threats and implementing these protective layers, the overall safety and reliability of the pipeline infrastructure can be greatly enhanced.

2.6.5 Prohibition of planting near transmission pipelines

All individuals and entities involved in land use and development activities must refrain from planting any vegetation within a minimum distance of 25 ft from any transmission pipeline. This precautionary measure is crucial not only to ensure the structural integrity of the pipelines but also to mitigate potential hazards that may arise from root systems, which could compromise pipeline safety.

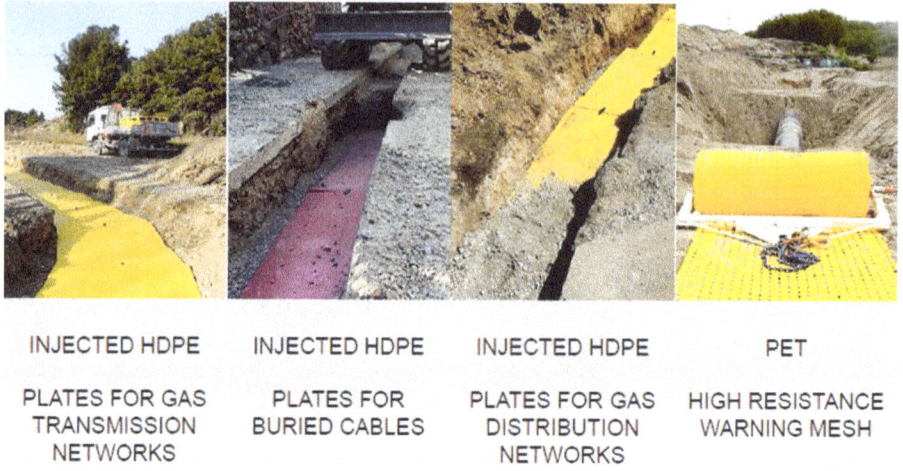

| INJECTED HDPE | INJECTED HDPE | INJECTED HDPE | PET |
| PLATES FOR GAS TRANSMISSION NETWORKS | PLATES FOR BURIED CABLES | PLATES FOR GAS DISTRIBUTION NETWORKS | HIGH RESISTANCE WARNING MESH |

Figure 2.9: Overpipe HDPE [3, 23].

Figure 2.10: Clearing of vegetation covering ROW markers and signs [3].

2.6.6 Strengthening community and landowner relations

To foster strong communication and collaboration, pipeline operators should actively engage with landowners and local communities. This engagement involves addressing land use concerns, providing transparent information about pipeline operations, and nurturing strong relationships through regular communication and feedback. By establishing trust and understanding, operators can better address any community issues that may arise, ultimately leading to more effective and responsible pipeline management.

2.6.7 Enhancing signage for onshore pipelines

To improve awareness of pipeline locations and promote safety, it is essential to enhance the signage along onshore pipelines (see Figure 2.11 and Figure 2.12). This includes the installation of clearly marked pipeline posts, standardized pipeline signs, indicators for road and railway crossings, as well as markers for riverbed protections and river crossings. Such signage is crucial to prevent unauthorized activities within the pipeline ROW and should be regularly inspected to ensure visibility and compliance with safety regulations (see Figure 2.10).

The signs installed should provide essential information such as
– The name of the pipeline owner
– A dedicated contact number for emergencies
– The type of product being transported through the pipeline
– The diameter of the pipeline
– Direction of flow
– The distance in kilometers from the nearest control station

This information will not only help inform the public but also serve as a warning to prevent accidental damage to the pipeline infrastructure. It is also important that a routine maintenance check on these markers occurs annually, ensuring that all paint and materials used during upkeep are nontoxic and safe for local livestock and wildlife.

2.6.8 Right-of-way intrusion detection system

To safeguard against unauthorized access or third-party interference, implementing an advanced intrusion detection system (IDS) or Threatscan system for aboveground piping is vital. This technology plays a critical role in quickly identifying disturbances and accurately pinpointing their locations along the pipeline.

For instance, the Guangdong Natural Gas Group, a leading Asian gas supplier, has successfully utilized fiber optic sensing as a means to detect leaks with remarkable

Figure 2.11: A marker post for a buried natural gas pipeline [3].

Figure 2.12: Examples of surface markers [3, 23].

precision. This system not only alerts personnel about possible threats in real time but also integrates seamlessly with Maxview Integration Software, providing a comprehensive monitoring solution that allows for proactive measures to be taken before any potential harm occurs.

2.6.9 Establishing a notification system

To enhance the safety and integrity of pipeline operations, the introduction of a structured third-party inquiry system is essential. This system could include initiatives such as "call before you dig" or a standardized "one-call system." Pipeline operators must develop and implement a documented strategy, often referred to as a damage prevention program. This program is aimed at minimizing risks of damage to pipelines and associated facilities during various excavation activities such as digging, trenching, and blasting. By establishing clear protocols and communication channels, operators can significantly reduce the potential for accidents and ensure a safer environment for both the pipeline infrastructure and the surrounding communities.

2.6.10 Safety management

The design regulations of most countries establish the essential precautions, safety assessments, and risk evaluations required to demonstrate the effectiveness of various safety measures such as minimum separation distances from buildings. The safety authority may require a risk analysis as a standard procedure to evaluate the level of risk associated with construction or excavation activities near pipelines. One notable safety management system employed by pipeline operators is the pipeline integrity management system (PIMS). It is strongly recommended that a comprehensive safety assessment or risk evaluation be conducted for pipelines situated in or near populated areas.

Key practices include
- Conducting annual evaluations of all potential threats by a multidisciplinary team.
- Reviewing emergency response and control plans based on the risk assessment findings.
- Strengthening administrative measures for ongoing monitoring of conditions.
- Implementing targeted initiatives focused on surveillance throughout the pipeline.
- If the class location has changed since commissioning, the following risk management strategies should be considered:
- Adjust the pipe segment to reflect the maximum allowable operating pressure (MAOP) appropriate for that class location.
- Reinforce the pipe by reinstating the MAOP to its original value.
- Increase the thickness of the existing pipe walls.

- Cut and replace sections with pipes that have a greater thickness or higher specified minimum yield strength (SMYS).
- Reassess the pipe segment to determine the new higher MAOP.
- Implement suitable risk mitigation measures informed by the risk assessment to
  ensure compliance with acceptable limits.
- Additionally, employ integrity assessment techniques as mandated by regulations
  and develop a performance plan dedicated to risk reduction.

2.6.11 Design factor

The concept of the design factor is essential in all pipeline regulations. The design factor refers to the ratio of the operating (hoop) stress in the pipeline to its yield stress.
With a few exceptions, the thickness of onshore pipelines is determined by hoop
stress in conjunction with the selected design factor, which limits the stress in the
pipe to a specific fraction of the SMYS of the material used. As illustrated in Table
2.14, design codes set boundaries for the design factor, allowing a maximum of 0.8 or
0.72 and a minimum of 0.4 or 0.3. In many instances, the lower design factors specified in the codes can be raised if additional safety measures are taken or if a risk and
safety analysis shows that the anticipated risk remains acceptable with a higher design
factor. Historically, and continuing today, third-party activities have been the primary
cause of significant failures in pipelines. Therefore, when there is an increased probability of third-party involvement, it necessitates a greater level of embedded safety
against such incidents. Utilizing a lower design factor or adopting a reduced design factor effectively provides this enhanced safety, as recognized by the codes [3, 22].

Table 2.14: General design factors [3, 24].

Location class ANSI B 31.8	Design factor B 31.8 (max)	Location class BS 8010	Design factor BS 8010 (max)
Class 1, Division 1	0.80	Class 1	0.72
Class 1, Division 2	0.72	Class 1	0.72
Class 2	0.60	Class 1	0.72
Class 3	0.50	Class 2	0.30–0.72[*1]
Class 4	0.40	Class 3	0.30[*2]

[*1]Variance to be justified by safety evaluation.
[*2]Maximum pressure limited to 7 barg.

In the context of the ASME B31.8 code, the application of a design factor of 0.8 is contingent upon specific testing requirements. When the maximum operating pressure
exceeds 72% of the SMYS of the pipeline material, a hydrostatic test must be conducted. This test involves pressurizing the pipeline with water to 1.25 times the desig-

nated design pressure, ensuring that the system can withstand higher pressure than it will typically encounter in operation.

Conversely, for pipelines located in other classification areas, there are alternative testing methods available. Instead of the traditional hydrostatic testing with water, these locations may permit the use of air or gas testing. This flexibility in testing approaches helps accommodate various operational contexts and environmental considerations associated with the pipeline infrastructure.

2.6.12 Mitigating damage to aboveground pipelines caused by third parties

Aboveground pipelines and their associated components face significant risks from various external threats, particularly those posed by vehicular collisions and vandalism. These risks can lead to serious damage, potential leaks, or hazardous situations. To effectively mitigate the likelihood of third-party damage to these aboveground systems, several strategies can be employed.

It is essential to consider enhanced protective barriers, such as guardrails and bollards, strategically installed around vulnerable sections of the pipeline. Regular surveillance and monitoring through CCTV or drone technology can help detect potential threats early on. Additionally, implementing clear signage to inform the public of the hazards associated with the pipelines can deter vandalism and unauthorized access. Engaging local communities and stakeholders in awareness programs may further strengthen the protective framework around these vital infrastructure components [3, 26].

1. Barrier-type prevention
–   Electrified fences: Ensure that electrified fences are installed properly and are maintained in good working condition. Regular inspections and checks for functionality are essential to guarantee their effectiveness in deterring unauthorized access.
–   Robust fencing and gates: Employ strong and durable fencing materials designed specifically to prevent unauthorized entry. This includes options such as barbed wire, antiscaling attachments, heavy-gauge wire, and thick wooden barriers. These materials create formidable obstacles for potential intruders.
–   Standard fencing options: Utilize conventional fencing like chain-link fences, supplemented with additional deterrents where necessary to enhance security.
–   Tamper-resistant locks: Install string locks and other locking mechanisms that are engineered to resist tampering and defeat, ensuring that unauthorized individuals cannot easily breach entry points.
–   Security personnel and canines: Deploy professional security guards who are trained and competent in asset protection or utilize trained guard dogs that are skilled in detection and apprehension of intruders.

- Alarm systems: Integrate high-decibel alarm systems as deterrents, which may include audible alarms and visual deterrents such as flashing lights designed to scare away intruders.
- Vehicle entry barriers: Implement barriers capable of preventing unauthorized vehicular access. In extreme cases, this might include employing ditches or other natural terrain features to act as obstacles. Physical barricades that require vehicles to navigate a slow and complex route can effectively deter rapid incursions.
- High visibility measures: Enhance surveillance by ensuring high visibility around the site. A well-lit and clear area makes it more difficult for potential intruders to approach the facility unnoticed.

2. Detection-type prevention
- Staffing levels: Maximize security effectiveness by maintaining full-time staffing levels with multiple personnel present at all times to provide surveillance and quick response capabilities.
- Video surveillance: Deploy advanced video surveillance systems that offer real-time monitoring and facilitate immediate response to incidents. Recording capabilities for later review are also crucial for investigations.
- Alarm systems with swift response: Utilize a range of alarms including sound monitors, motion sensors, and comprehensive alarm systems. Quick and decisive responses to alarms are vital to thwart potential breaches.
- Supervisory Control and Data Acquisition (SCADA) systems: Implement SCADA systems that can immediately indicate tampering with critical equipment. Changes in signals sent to the control room can serve as alerts to imminent threats.
- Satellite surveillance: Take advantage of satellite technology, which now offers high-resolution capabilities for continuous monitoring of pipelines and surrounding areas, ensuring prompt detection of unusual activities.
- Explosive dye markers: Install explosive dye markers designed to spray a dye on intruders during a breach, aiding in their identification, apprehension, and prosecution efforts.
- Comprehensive IDSs: Integrate sophisticated IDSs that provide alerts in real time. This is essential for timely intervention to prevent security incidents.

All detection-type prevention measures must be paired with prompt responses to enhance their efficacy.

3. Patrolling
- Dynamic patrol schedules: Vary patrol and inspection schedules frequently to keep potential saboteurs guessing. Unpredictable patrol patterns are effective in deterring malicious acts by reducing the chances for planning by would-be intruders.

4. Simulated security measures
- Decoy deterrents: Utilize simulated measures such as plastic imitations of steel bars, fake surveillance cameras, and warning signs that imply the presence of security measures that are not actually in place. While these may not provide as strong a deterrent as real security features, they can still create an impression of security that might dissuade some intruders.

2.7 Improper operations mitigation

If a thorough evaluation of failures or risk assessments reveals a significant vulnerability to operational deficiencies, it is recommended that one or more of the following risk mitigation strategies be implemented to enhance the overall safety and efficiency of operations:
- Review operator training and qualification programs: Conduct a comprehensive analysis of the existing training protocols to ensure that operators possess the necessary skills and knowledge. This may include revising training materials, increasing hands-on training sessions, and implementing assessment tools to measure competency levels.
- Review standard operating procedures (SOPs) and the operations and maintenance manual: Systematically evaluate the current SOPs and the operations and maintenance manual to identify any areas that require updates or clarifications. This process should involve collaboration with frontline employees to ensure that procedures reflect actual practices and address any ambiguities that could lead to operational failures.
- Assess operating control equipment (SCADA, ESD) for enhanced process controls: Perform a detailed assessment of the existing SCADA systems and emergency shutdown (ESD) systems to determine their effectiveness in monitoring and controlling processes. This may include upgrading software, improving sensor accuracy, and enhancing alarm systems to prevent control failures.
- Implement robust operating procedures: Develop and enforce comprehensive and clear operating procedures that guide operators through critical tasks. These procedures should include detailed step-by-step instructions, risk assessments, and contingencies for various scenarios to minimize the likelihood of errors during operations.
- Conduct an independent audit: Engage a qualified third-party auditor to carry out an impartial review of operational practices, training programs, and safety protocols. The audit should provide an objective assessment of vulnerabilities and recommend specific actions to address any identified weaknesses.

By systematically addressing these areas, organizations can significantly reduce their susceptibility to operational deficiencies and enhance their overall safety culture.

2.8 Mitigation from external forces

During routine ROW patrolling, if any risks related to ground movement, soil erosion, or the scouring of river and creek bottoms are identified, a series of targeted risk mitigation activities should be promptly initiated to ensure the continued integrity of the pipeline. The following detailed steps should be undertaken:

1. Depth of cover and elevation survey
   - Conduct a thorough depth of cover survey for the affected section of the pipeline to measure how deep the pipeline is buried.
   - Perform an elevation survey to ascertain the topography of the surrounding land, ensuring we understand how elevation changes may impact the pipeline's stability.
2. Assessment of underground movement
   - Utilize advanced monitoring equipment, such as inclinometers, to detect any tilting or shifts in the ground that could indicate potential ground movement.
   - Deploy strain gauges along the pipeline to monitor for any unusual stress or strain that may result from external geological forces.
3. Hydro technical and geotechnical engineering evaluations
   Engage qualified engineers to conduct a comprehensive assessment of the site, focusing on both hydrotechnical and geotechnical factors to evaluate the current conditions and potential risks to the pipeline.
   - The evaluation will determine suitable remedial action options, which may include:
     - Pipeline rerouting or replacement:
       - If significant risks are identified, consider pipeline rerouting or replacement using horizontal directional drilling techniques to bypass the affected areas safely.
     - Line lowering within the existing ROW:
       - Assess the feasibility of lowering the pipeline within its current ROW to a more stable depth, thus reducing exposure to potential ground movement or erosion hazards.
     - Armoring of approach slopes and banks:
       - Implement armoring techniques on the approach slopes and banks of rivers or creeks to stabilize the soil and prevent further erosion, thereby protecting the integrity of the pipeline from environmental forces.

By systematically following these detailed steps, we can effectively manage and mitigate risks to the pipeline infrastructure and maintain operational safety.

2.9 Mitigation of seismic and landslide hazard

Key ingredients of pipeline seismic hazard mitigation include
– geohazard assessment: geohazard identification and characterization,
– pipe-soil interaction analysis,
– strain-based design criteria, and
– weld strength and quality.

For seismic loads, examination of papers and design guidelines show that for a welded steel pipeline, ground shaking and strain do not cause any substantive harm. Problems for buried pipelines exist either where a fault with a surface expression is crossed or where the ground in which the pipeline is laid loses strength and a *land slide or slip leaves* the pipe without support or imposes large shear forces when moving. The fault movement can be designed with long straight lengths of pipe crossing the fault at an angle of 60° and the pipeline trench backfilled with cohesionless backfill (e.g., gravel). The following mitigation strategies are suggested [3].

2.9.1 Geohazard assessment

Identification of potential *landslide* areas should form part of the route survey in seismically active areas and susceptible areas avoided wherever possible.

Geohazard assessment provides information on the earth's surface processes and geological hazards that might pose a threat to pipelines and associated infrastructure. Geohazard assessment of the project site should cover the following:
a) Desk study
    Perform a desk study to describe the geologic setting along the alignment and conduct a preliminary assessment of the Geohazard. The main inputs to the desktop study include aerial photography, geologic maps from government agencies and local authorities, as well as data from open sources (i.e., province boundaries, district boundaries, roads, and rivers).
b) Mapped earthquake faults and seismic zones
    – Earthquake vibrations
    – Ground liquefaction including lateral spreading
    – Fault rupture
    – Tsunami
c) Intensity of ground shaking at the project site determined by probabilistic methods (10% probability of occurrence in 50 years)
    – Ground settlement and collapse: Swelling, shrinking ground and collapsible soils, ground solution, and collapse (karst)

d) Potential for liquefaction, ground failure, and landslides at the site
   - Landslides and erosion: Pre-existing landslides and erosion areas, landslide-prone, and erosion-prone terrain
e) Potential for flooding at the site from man-made facilities and natural storms
   - Coastal erosion and deposition
   - River behavior including channel changes and bed erosion
   - High water table and flooding
f) Contaminated land and ground geochemistry of the project site including methods for mitigation:
   - Naturally aggressive ground and groundwater
   - Former and current industrial uses
   - Contamination of soil and rock
   - Contamination of groundwater and surface water

2.9.2 Re-routing

Pipelines have been built in seismically active areas both as buried and surface run pipelines. In general, there is little advantage to having the pipeline aboveground as general shaking is not a particular problem for buried pipelines, but pipelines have been known to fall off supports or be excessively strained at individual supports, producing a point load, not associated with buried pipelines [3].

In the event of mass ground failure, both types of construction are equally vulnerable and the pipeline would be fortunate to survive. Where other considerations apply, supports have been specifically designed to move only under large forces which would be close to the yield strength of the pipeline. Project examples are rare as particular fault areas are normally avoided wherever possible and where ground failure is predicted from soil analysis, rerouting is normally undertaken.

2.9.3 Pipe–soil interaction analysis

The prediction and modeling of soil pipe interaction is a key driver in the design of many deep-water and high-temperature/high-pressure (HT/HP) pipeline systems. Design of pipelines to allow lateral buckling or prevent upheaval buckling and the design of expansion spools all require a detailed knowledge of pipe-soil interaction [3].

In recent times many soil pipe interaction tests have been performed and some guidelines have been produced, for example, hotpipe and safe-buck.

2.9.4 Strain-based design

The design based on strain is suitable when the stresses and strains go beyond the proportional limit, and when the maximum design loads should be decreased as the material strains.

Strain-based design has been found to be useful for laying offshore pipes, operating pipelines at high temperatures, pipelines in areas with soil movement, and arctic pipelines.

Soil movement should generally be considered as controlled by displacement. However, there are known situations where soil-induced loads are controlled by load or are intermediate between load and displacement control [2].

It is generally not practical to design pipeline crossings of permanent ground displacement (PGD) zones (i.e., high-strain conditions) in accordance with the usual code-allowable stresses for operating and external load conditions. Instead, the accepted alternative approach is to revert to a strain-based design approach that allows comparatively large strains to occur locally in the pipeline provided that pressure boundary integrity is assured – that is, allow damage to the pipe but prevent pipe rupture and release of contents. Tensile strain limits may be taken as high as 3–4%, depending on the quality of the girth welds. Compressive strain limits typically range from about 2% to 4% for range of diameter-to-thickness ratio (D/t) of 90 down to 45. Strain acceptance limits assume that the pipeline girth welds should be capable of developing gross section yielding of the pipe wall. This capability, often referred to as "overmatching welds," means that failure would occur in the pipe before failure in the weld or the weld heat-affected zone. This implies that a welding process and weld inspection program to minimize both the number and size of imperfections would be implemented during construction. The principles of strain-based design are well-chronicled in the technical literature and should not be repeated here. Likewise, the methodology for performing nonlinear finite element analysis of pipelines subjected to PGD is well-established and described in available industry guidance documents.

For active faults and liquefaction zones, a pipeline must be designed to withstand the full effect of the PGD without rupture, as there is no time for intervention once the event is initiated [3].

For landslides, knowing the limitations of strain-based design for landslide hazard mitigation, it is essential that a monitoring program be established to detect ground movements prior to reaching damaging levels, which might vary among respective sites. Ideally, the extent and scope of monitoring should be based on site-specific risk assessments of the inventory of landslide areas, subject to adjustment based on the findings and conclusions from periodic observations and evaluation of data [3].

2.9.5 Scheduling

To effectively reduce the risk of pipeline routing through geohazard areas, it is essential for pipeline projects to incorporate thorough geological investigations at the earliest stage of the front-end engineering design process. This proactive approach allows for the identification and assessment of potential geohazards – such as landslides, sinkholes, and fault lines – before finalizing the pipeline route. By prioritizing these investigations, project teams can make informed decisions regarding the alignment and design of the pipeline, ultimately enhancing safety and minimizing environmental impacts. Additionally, early geological assessments can lead to cost savings by avoiding later modifications or disruptions during construction due to unforeseen geological challenges.

2.10 Leak and break detection

Pipeline operating companies are required by Clause 14.124 of LI 2189 to make periodic pipeline balance measurements to check system integrity [1]. Both installed devices and operational procedures must be in place to detect pipeline failures early.

Operations personnel must be diligent in the observation of pipeline and pipeline system components during field surveillance. Knowledge of normal operating conditions, such as system pressures, is integral to leak detection. Not all pipeline leaks are noticeable by operating conditions. Therefore, during daily rounds, the operator must observe line and lease conditions that may result in a failure [3].

Production volume discrepancies must be considered daily since low production for no apparent reason may signify a pipeline leak or rupture.

In the event of a report of a problem or spill from the public or another outside party, the operator must immediately investigate.

If a pipeline leak or rupture is detected, the source of the released product must be isolated immediately. If there are multiple possibilities, isolate all possible sources and determine the correct source after the release is under control.

2.10.1 Leak detection methods

Leak detection is used to determine where a leak has occurred in liquid and gas pipeline systems. Methods of detection include hydrostatic testing (hydrotest), infrared, and laser technology after pipeline erection and leak detection during service [3].

Causes of pipeline leakage can be divided into five main categories:
- Internal and external corrosion
- Third-party damage
- Operational error

– Natural hazards
– Mechanical failure

Leak detection for the pipeline may be achieved by the provision of equipment (e.g., PIMS) to undertake constant monitoring of pressure sensing devices located at each end of the pipeline and at the intermediate isolating valve stations.

The output from these monitoring devices should be displayed in the main control room, thereby enabling the operators to identify abnormal or unexplained deviations in pressure and to shut in the pipeline section affected by actuating the intermediate/block isolating valves.

In addition to the above, any PIMS/SCADA system, metering should be installed at each end of the pipeline. The signal from the meter at the receipt terminal should be transmitted to the main control room (usually the pumping end of the pipeline) for comparison with the outgoing meter signal. Any unexplained deviation from a predetermined threshold value should alert the operators as to a possible leak or malfunction [3].

A leak detection system may be classified as

a. Internal-based leak detection system
b. External-based leak detection system

External-based leak detection systems can detect the smallest leak with high accuracy. Internal-based leak detection discovers gas leakage based on measurement readings at some specific location along the pipeline, for example, computational pipeline monitoring (CPM).

The main aim of leak detection is to help pipeline operators to detect and localize leaks. The following methods may be used to detect leaks in pipelines.

Table 2.15 describes common techniques that should be considered for detection of pipeline leaks.

Table 2.15: Summary of some leak detection techniques [3].

Technique	Description
Soap solution bubble test	The pressurized unit designated for testing is thoroughly coated with a specially formulated soap solution. This solution creates a slick, bubbly layer on the surface of the unit. As the unit undergoes pressurization, any escaping gas will interact with the soap solution, forming visible bubbles at the point of the leak. The operator closely observes the unit, looking for these bubbles, as they indicate the specific areas where gas is seeping out. This method provides a clear visual cue to identify leaks that need to be addressed for safety and functionality.

Table 2.15 (continued)

Technique	Description
The water immersion bubble test method	The water-immersion bubble test, often referred to as "bubble testing" or "dunking," is a traditional and somewhat rudimentary method for detecting leaks. This technique involves submerging a charged or pressurized component, typically filled with high-pressure dry air or nitrogen, in a water tank and observing for any escaping bubbles.
Software-based leak detection system	Computational pipeline monitoring (CPM) systems, also known as software-based leak detection systems, utilize pipeline data to identify potential leaks and to alert operators to hydraulic anomalies that exhibit characteristics typical of a leak. These systems are designed to notify the pipeline controller, enabling them to assess the situation and, if necessary, shutdown the pipeline to minimize the impact of any spill.
H_2S detection	– Employ either permanent or portable detection tools to monitor for hydrogen sulfide (H_2S) gas concentrations effectively in various environments. – Permanent gas monitoring systems are typically installed at surface facilities to provide continuous and reliable surveillance of H_2S levels, ensuring that any dangerous buildup is detected promptly to maintain safety.
Right-of-way (ROW) surveillance	Conducting a visual inspection through walking or aerial surveillance can help identify potential leakages along a natural gas pipeline. Signs that may indicate a suspected gas pipeline leak include – Soil settlement, the presence of gas bubbles, and discoloration of water, soil, or vegetation – Whistling or hissing sounds – A distinctive, strong odor often likened to that of rotten eggs – Dense fog, mist, or white clouds – Bubbling in water bodies such as ponds or creeks – Dust or dirt being expelled from the ground – Discolored or dead vegetation located above the pipeline ROW These methods can be effectively combined with infrared thermography and flame ionization surveys for enhanced detection.
Production monitoring	– Volume balancing (inlet and outlet flow) or pressure monitoring (pressure drop) to detect gas leakage – Changes in production volumes or pressure may indicate pipeline failure – It is a more effective tool for finding large leaks and cracks
Detecting leaks through pressure changes	A leak can cause a significant change in gas pressure. Therefore, sensors can be installed to detect pressure changes in the pipeline. Changes in pressure can trigger an alarm. The sensors required for this technique can be categorized as flow, pressure, and temperature.

Table 2.15 (continued)

Technique	Description
Fiber optic sensing technology	Leaks in utility pipes can lead to abrupt variations in temperature within the soil that encases them. This is significant because fiber optic cables, strategically installed alongside these pipes, are capable of detecting these thermal fluctuations. Additionally, these advanced cables can pick up on acoustic vibrations generated by the leaking pipes. When a leak occurs, the combination of temperature shifts and sound changes sends signals through the fiber optic system to a centralized control room. If any of these signals indicate an irregularity or an anomaly in the expected conditions, it triggers an alarm, allowing for prompt investigation and remediation of the issue. This technology enhances the ability to monitor the integrity of pipeline infrastructure in real time, improving both safety and maintenance efforts.
PIMS/SCADA-based system	The leak detection system is designed for seamless integration with the SCADA architecture in the control room. By leveraging SCADA capabilities, the system can provide real-time alerts to operators upon leak detection and maintain comprehensive logs for both historical analysis and trend monitoring pre- and postevent. Precise geolocation of the leak enhances the efficiency of response protocols, enabling quicker interventions and minimizing downtime during repairs.
Flame ionization survey	– Utilizes advanced electronic instrumentation for the detection of trace gas concentrations at extremely low levels. – The device features portable and highly sensitive capabilities, necessitating the repositioning of piping to target flammable gas sources. – Configuration options include hand-held operation, ATV-mounted deployment, or installation on rotary-wing aircraft for aerial surveillance.
Infrared thermography	– Thermal imaging technology effectively identifies temperature variations on the right side, indicative of gas leaks or produced water discharges. – A critical factor is ensuring a sufficient volume of gas escape, which is necessary to generate a detectable thermal differential. – This detection is typically accomplished through aerial surveillance methods.
Odor detection	The process of gas odor detection involves utilizing specially trained animals, along with the application of patented odorants. These trained animals, known for their keen sense of smell, are capable of identifying pinhole gas leaks that would typically go unnoticed through conventional detection methods. This innovative approach not only enhances safety by ensuring even the smallest leaks are detected but also demonstrates the effectiveness of combining biological capabilities with technological advancements in leak detection systems. By integrating these trained animals into monitoring protocols, we can significantly improve our ability to identify and address potential hazards posed by undetectable gas leaks in various environments.

Table 2.15 (continued)

Technique	Description
External leak detection equipment	An external leak detection device can be strategically installed along the pipeline to enhance safety and monitoring capabilities. This advanced detection system continuously analyzes the flow dynamics of the pipeline, assessing various parameters such as pressure, temperature, and flow rate. By measuring these factors in real time, the device can identify subtle changes or irregularities that may suggest the presence of a leak. Additionally, it can differentiate between normal operational variations and unusual anomalies, allowing for prompt detection and intervention to prevent potential environmental damage or safety hazards.
Intelligent pigging	Small leaks in pipelines can generate ultrasonic signals, which are capable of being detected by a specialized device known as a "smart pig." This instrument is propelled through the pipeline by the flow of oil, allowing it to traverse long distances in a short amount of time – often within mere seconds. As a result, this technique can capture several hundred samples during a single pass, making it highly efficient for monitoring the integrity of the pipeline. One of the key advantages of utilizing this method is its sensitivity; it can identify very small leaks that might otherwise go unnoticed, helping to prevent more significant issues from arising. However, a notable disadvantage of this approach is the frequent requirement for the deployment of pigs. Each inspection necessitates the use of these devices, which can lead to operational challenges and increased costs associated with their maintenance and handling.
Acoustic emission systems	The indication of a leak often becomes evident through an elevated noise level. This increase in sound intensity is typically a direct result of the escaping fluid, whether it be gas or liquid, creating turbulence and vibrations as it moves through or around obstacles. The specific audio signature produced by the leak can serve as a crucial tool for both detection and localization. By analyzing the characteristics of the noise – such as its pitch, frequency, and volume – technicians and engineers can pinpoint the location of the leak with greater accuracy, allowing for timely repair and mitigation of any potential damage or hazards.
Radioactive tracing	This advanced technology is extensively utilized for leak detection in process piping and heat exchanger systems, particularly those that transport liquids and gases within the oil and gas industry. The process begins by introducing a small quantity of radioactive material at one end of the pipeline. This isotope is specifically chosen for its distinct radiation signature, which enables effective tracking. To monitor the movement of this radioactive tracer, a radiation detector is strategically positioned outside the pipeline, or in some cases, above ground. As the tracer travels through the pipe, the detector continuously measures the levels of radiation detected.

Table 2.15 (continued)

Technique	Description
	The detection process hinges on observing the radiation levels as the tracer moves. A significant reduction in detected radiation signals the presence of a leak or blockage within the system. This decrease occurs because the radioactive material escapes into the surrounding environment in the case of a leak, or it is obstructed by a blockage, preventing it from reaching the detector. This method not only helps identify the location of leaks but also provides crucial information that can assist in maintaining the integrity and safety of critical piping systems in the industry.

2.10.1.1 Leak detection by mass balance

A pipeline leak is characterized by the unintended release of the material being transported through the pipeline system, which can involve liquids, gases, or slurries. To effectively identify such leaks, the most comprehensive and scientifically sound approach employed by online pipeline leak detection systems is known as "mass balance." This technique involves precisely calculating the discrepancy in the amount of material entering and exiting the pipeline, allowing for the detection of any loss in transit. By measuring the total mass of the substance at various points along the pipeline and comparing the figures, the mass balance method provides valuable insights into the efficiency and integrity of the transportation infrastructure, specifically expressed in terms of mass units. This rigorous assessment is crucial for ensuring the safety and reliability of pipeline operations as well as for minimizing environmental impacts and operational losses:

$$\Delta M(t) = M_{\text{in}}(t) - M_{\text{out}}(t) - M_i(t) \tag{2.1}$$

where $\Delta M(t)$ is the pipeline leakage as a function of time t, $M_{\text{in}}(t)$ the measured flow into the pipeline, $M_{\text{out}}(t)$, measured flow out of the pipeline, and $M_i(t)$ the pipeline inventory (accumulation or amount of fluid determined to be contained in the pipeline).

The following formula calculates natural gas's estimated line pack volume [8]:

$$LP = \left(\frac{T_b}{T_{\text{av}}}\right) \times \left(\frac{P_{\text{av}}}{P_b}\right) \times V_{\text{seg}} \times F_{\text{pv}}^2 \times C_f \tag{2.2}$$

where LP is the calculated natural gas volume (in MCF) in a pipeline segment; the line pack [27]; T_b the base temperature (in degrees Rankine) for line pack. The base temperature is specified as a system input (not per pack segment) but is user-configurable. The default value is 60 °F or 519.67 °R; T_{av} the average flowing temperature (in degrees Rankine) in the pipeline segment; P_{av} the average pressure (in psia) in the pipeline segment;

Table 2.16: Summary of hazard mitigation options [3].

Threat	Description	Integrity assessment	Mitigation	Interval – years
Improper design or materials selection	If failures, direct examinations, or thorough risk assessments reveal a significant vulnerability to unacceptable design or materials, it is essential to implement one or more of the following risk mitigation actions. These actions may include conducting comprehensive redesign efforts to enhance structural integrity, sourcing alternative materials that meet higher safety standards, implementing stringent manufacturing quality control procedures, or increasing the frequency of inspections and evaluations throughout the lifecycle of the project. Additionally, involving cross-functional teams to assess the potential risks and develop tailored strategies for improvement can be crucial. Ensuring stakeholder engagement and continuous monitoring will further strengthen the overall approach to mitigating identified risks.		– Conduct an in-line inspection of the pipeline to assess the specific design deficiencies and material anomalies. – Perform pressure testing on the pipeline to detect any weld defects or material inconsistencies.	

Table 2.16 (continued)

Threat	Description	Integrity assessment	Mitigation	Interval – years
As-built flaws mitigation	Design and manufacturing flaws can arise from various issues throughout the production and construction processes. These may include defects in the fabrication of the steel such as impurities or inconsistencies that affect the material's overall integrity. Additionally, out-of-specification material properties can occur when the steel does not meet the required standards for strength, durability, or corrosion resistance. Moreover, flaws can also stem from the construction process itself, where improper assembly or installation techniques may compromise the structure's stability and performance. Finally, problems may arise in the protective coatings applied to the steel, which are essential for preventing corrosion, or in the cathodic protection (CP) systems designed to enhance the lifespan of the structure. Each of these factors could significantly impact the reliability and safety of the final product.	Conduct hydrotest	– Fabrication specifications: Detailed guidelines outlining the materials, processes, and quality standards required for the manufacturing of components. This includes specific tolerances, workmanship requirements, and testing protocols to ensure that the fabricated items meet predetermined performance criteria. – Construction specifications: Comprehensive instructions and standards governing the construction processes. This document encompasses site preparation, installation procedures, safety measures, and material handling requirements, ensuring that construction activities adhere to relevant codes and industry best practices.	N/A

Table 2.16 (continued)

Threat	Description	Integrity assessment	Mitigation	Interval – years
			– Independent verification: A process in which an unbiased third party evaluates and confirms that the fabrication and construction processes comply with all specified requirements. This verification includes examining documentation, conducting inspections, and performing tests to ensure the integrity and quality of the work.	
			– Independent audit: A thorough examination conducted by an external entity to assess compliance with established project standards, regulations, and contractual obligations. The audit includes a review of records, interviews with personnel, and onsite inspections to identify areas of noncompliance or opportunities for improvement.	

Table 2.16 (continued)

Threat	Description	Integrity assessment	Mitigation	Interval – years
			– Pressure hydrotest: A critical procedure used to evaluate the strength and integrity of piping systems. This testing involves filling the system with water and subjecting it to predetermined pressure levels for a specified duration to ensure that there are no leaks or weaknesses in the structure. – Replace pipe at test failure locations: The necessary action to identify and remove sections of piping that fail during testing procedures. This step includes evaluating the extent of failure, determining appropriate replacement materials and methods, and retesting the system to verify that all repairs meet the required standards for safety and performance.	

Table 2.16 (continued)

Threat	Description	Integrity assessment	Mitigation	Interval – years
Corrosion	Pipelines, whether subterranean or elevated, are subjected to corrosive environments from both internal and external sources.	– Execute hydrostatic testing – Carry out in-line inspections – Implement external corrosion direct assessment (ECDA)	Internal corrosion: Process control, for example, dehydration, gas quality specification Corrosion inhibition Regular Cleaning Pigging External corrosion: Cathodic protection Coating Stress corrosion cracking (SCC): Replace pipe at test failure locations	External and internal corrosion: Ten (10) years For SCC: 3–5 years
Third-party damage	The threat of third-party accidental or deliberate impact on the pipeline is an event-based scenario.		– Regular survey – Public awareness – Depth of burial	
Improper operations	Operational misuse arises as a failure event through incorrect operating procedures or a failure to follow the correct procedure by personnel.	None required	Review operator training and qualification programs. Review standard operating procedures and pipeline operations and maintenance manual. Assess operating control equipment (SCADA, ESD) to improve process controls. Robust operating procedures Independent audit	N/A

P_b the base pressure (in psia) for line pack. The base pressure is specified as a system input (not per pack segment) but is user-configurable. The default value is 14.73 PSIA; V_{seg} the volume (in MCF) of the pipeline segment based either on the segment's configured length and diameter or a user-configured volume; F_{pv} the super compressibility factor for the pipeline segment. The super-compressibility factor is user-configurable for each pipeline segment; and C_f the volume correction (conversion) factor. The correction factor is user-configurable for each pipeline segment.

The gas compressibility factor, Z_a, accounts for all deviations of a real gas from ideal gas behavior, thus

$$Z_a = \frac{\text{Actual volume of gas}}{\text{Ideal volume of gas}} \tag{2.3}$$

or for the general equation of state $P = Z_a NRT$.

The gas compressibility factor (Z_a) for the gas at average conditions is calculated based on the California Natural Gas Association (CNGA) equation:

$$Z_a = \frac{1}{1 + \frac{\left(C_{Z_a} \times 10^{1.785G} \times \bar{P}_g\right)}{T_a^{3.825}}} \tag{2.4}$$

where C_{Z_a} = a constant 3.444×10^5 for the f.p.s. units stated or 5.172×10^5 for c.g.s. units. Therefore,

$$Z_a = \frac{1}{1 + \frac{\left[\left(3.444 \times 10^5\right) \times 10^{1.785G} \times \bar{P}_g\right]}{T_a^{3.825}}} \tag{2.5}$$

Data required for the analysis are gas gravity, G (air = 1), average pressure (psig), and average temperature (°C).

2.11 Strategies to reduce stress corrosion cracking in pipelines

2.11.1 Material selection

The first line of defense in controlling stress corrosion cracking is to be aware of this possibility at the design and construction stages. By choosing a material that is not susceptible to self-compacting concrete in the service environment, and by properly processing and manufacturing it, subsequent self-compacting concrete problems can be avoided. Unfortunately, it's not always that simple. Some environments, such as high-temperature water, are very aggressive and should cause self-compaction in most materials. Mechanical requirements such as high yield strength can be very difficult to reconcile with SCC resistance (especially where hydrogen embrittlement is involved). The material selection process should reflect the overall philosophy regarding design life, cost profile, inspection and maintenance philosophy, safety and environmental profile, failure risk assessment, and other specific project requirements [28].

The material selected for the acidic operating environment should be resistant to SCC and meet the requirements of the following standards:

– CSA Z662, clause 16, Sour service
– NACE MR0175/ISO 15156 for upstream oil and gas process piping within the scope of ASME B31.3 – materials for use in H_2S-containing environments in oil and gas production
– NACE MR0103 for downstream refining and upgrading process piping within the scope of ASME B31.3
– NACE TM 0284–96, evaluation of pipeline and pressure vessel steels for resistance to hydrogen-induced cracking (HIC)

NACE MR0175/ISO 15156 is an international standard that provides requirements for metallic materials exposed to H_2S in oil and gas production environments. NACE MR0175 defines sour service as "Exposure to oilfield environments that contain sufficient H_2S to cause cracking of materials by the mechanisms addressed by NACE MR0175/ISO 15156." Hardness and chemical composition control are the basis for mitigating H_2S corrosion, which can form in various forms such as sulfide stress cracking corrosion, stress cracking corrosion (SCC), HIC corrosion, hydrogen-induced stress cracking corrosion, and corrosion. The NACE MR0175/ISO 15156 standard addresses all cracking mechanisms:

– Sulfide stress cracking (H_2S)
– Stress corrosion cracking (Cl)
– Environmental cracking (synergistic, H_2S, and Cl)
– Various HIC mechanisms

2.11.2 Environment

The most direct way to check for SCC using an environment scan is to remove or replace the environment component responsible for the problem, although this is usually not possible. Where crackers are required as environmental components, environmental control options include adding inhibitors, modifying the electrode potential of the metal, or isolating the metal from the environment with coatings. Corrosivity of H_2S is a major risk for carbon steel, austenitic stainless steel, and other CRA materials. Therefore, a gas sweetening process unit should be designed and considered for gas processing plants with sour operation [3].

2.11.2.1 Gas conditioning systems

Natural gas, whether produced from a condensate field or as associated gas from an oil field, usually contains water vapor (H_2O) and often contains H_2S and/or CO_2 and heavy hydrocarbons. Other contaminants in the gas include CS_2, COS, mercaptans (RSH), N_2, O_2, Hg, solid hydrates, asphaltenes, and dust. Some or all of these impurities must be removed to meet gas specifications. This is done in the gas treatment module. The gas treatment module is usually installed at the entrance to the gas treatment plant [3].

2.11.2.1.1 Inlet separation

Inlet separation removes any remaining water and heavy hydrocarbons from the gas stream.

2.11.2.1.2 Gas treating (sweetening)

Gas conditioning is used to reduce the "sour gases" of carbon dioxide (CO_2) and hydrogen sulfide (H_2S), along with other sulfur-containing compounds, to low enough levels to meet contract specifications or to allow postprocessing at the plant without corrosion and problems with socket [3].

One of the most dangerous compounds present in natural gas is H_2S. Its removal is necessary to limit corrosion. H_2S gas is highly toxic. In addition, CO_2 creates many technical and safety problems if the concentration exceeds the specified limit according to gas sales quality requirements. CO_2 forms a strong acid that is highly corrosive in the presence of water. Carbon dioxide is also nonflammable, and as a result, large amounts reduce the calorific value of the fuel.

The removal of H_2S and CO_2 from natural gas is called "gas and liquid sweetening." Many such sweetening technologies have been developed. H_2S and CO_2 can be removed from gas streams by physical absorption, chemical reaction and adsorption, or a combination of these processes. A common way to remove H_2S and CO_2 from natural gas is to use a solvent-amine chemical system that uses a stepped contact tower or structured packing to pass the acid gas through the amine liquid, absorbing the H_2S and some of the CO_2. The principle of this process is shown in Figure 2.13.

2.11.2.1.3 Other conditioning process

Selective removal of free gas such as N_2, He, Ar, and Ne from natural gas is done using a special type of molecular sieve.

When using aluminum heat exchangers and equipment, it is often necessary to remove mercury from the gas. This is usually done by passing the gas through a layer of sulfur-impregnated activated carbon or alumina where the mercury reacts to form mercuric sulfide, H_2S.

Filters are used to remove solid particles.

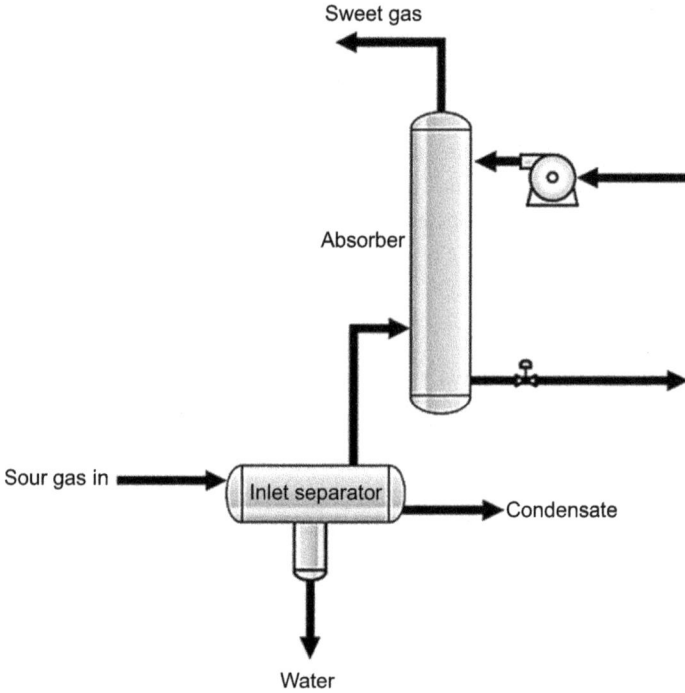

Figure 2.13: Typical amine gas treatment [3].

2.11.3 Stress

Pipe stress can lead to premature degradation of pipe strength. Stresses acting on the pipe include:
– residual stress from the manufacturing process;
– external stress such as stress from bending, welding, mechanical scratching, and corrosion; and
– secondary stress due to settlement or soil movement.

One of the causes of stress corrosion cracking is the presence of stress in piping components, so to mitigate SCC you must eliminate stress or reduce it below the threshold stress for SCC. If SCC occurs during welding or forming, then use stress relief annealing to relieve the stress [3, 29].

Partial stress relief around welds and other critical areas can be valuable for large structures where complete stress relief is difficult or impossible. Stress can also be alleviated mechanically. For example, hydrostatic testing beyond bearing capacity should tend to "even out" the stress and reduce the maximum residual stress.

Laser blasting shot peening or blasting can be used to introduce a surface compressive stress that is beneficial for SCC control. These processes can be performed uniformly.

2.11.4 Coating

Clean and prepare tube surfaces by shot peening (shot peening is a cold work process used to finish the surface of metal parts to prevent fatigue and stress corrosion damage and extend the life of the shot peened part (round metal, glass, or ceramic particles) with sufficient force to produce plastic deformation and then apply special coatings such as fusible epoxy to protect the pipe from SCC [3, 29].

2.11.5 Cathodic protection

Use CP systems to protect pipes from corrosion, carry out annual surveys to check the effectiveness of CP, and repair areas with pipe-to-ground potential below 0.85 V [3, 5, 30].

Methods to prevent corrosion and SCC on existing pipelines include minimizing the operating temperature and controlling CP levels to values more negative than 850 mV CSE. Minimizing pressure fluctuations in service pipelines is also effective in preventing SCC initiation [29].

2.12 Hydrate mitigation

High pressure and low temperature are the most favorable circumstances for hydrate formation. For pipeline systems, the lowest temperature usually occurs during shutdown, hence this circumstance should always be considered. Free water must be present in the pipeline or processing system [3].

Therefore, conditions that are known to encourage hydrate formation are:

a. Low temperature
   - Hydrate forms at low temperature called formation temperature. If a gas-containing water is cooled below its hydrate formation temperature, hydrates will form

b. High pressure
   - High pressure favors hydrate formation

c. Water
   - Free water or water vapor at sufficiently low temperature or high pressure enhances hydrate formation. No hydrate formation is possible if "free" water is not present. Here, we understand the importance of removal of water vapor from natural gas so that in case of free water occurrence there is likelihood of hydrate formation.

Further conditions that are known to promote hydrate formation are
– high velocities;
– pressure pulsations (in other words, turbulence can serve as a catalyst);
– agitation;
– The presence of H_2S and CO_2 promotes hydrate formation because both these acid gases are more soluble in water than the hydrocarbons.

It is important to note that
– high BTU gas is more likely to produce hydrates and freezing problems;
– Joule-Thomson rule of temperature effect because of pressure reduction. Temperature will decrease by approximately 7 °F for every 100-psi pressure reduction.

The hydrate crystals forming on pipe walls initiate any unevenness such as weld seams or areas of corrosion. Narrowing of the internal pipe diameter starts slowly, but as it changes flow and pressure conditions, hydrate build-up is accelerated. Natural gas hydrate will easily form at
– elbow;
– weld seams;
– areas of corrosion;
– at the downstream of the valve there is a big pressure drop;
– relief valve and the safety valve of the high-pressure vessel; and
– pig launcher, pig receiver.

2.12.1 Methods of hydrate prevention and mitigation

To prevent the formation of hydrates within the process the following methods are used.
1. The temperature is maintained above the hydrate formation temperature of the gas. Increasing the gas temperature above the temperature indicated in the hydrate formation curve for the given operating pressure.
2. The pressure is maintained below the value necessary for hydrates to form. Decreasing the pressure below the pressure indicated on the hydrate formation curve for the given operating temperature.
3. Following the recommended chemical injection rates chemicals called hydrate inhibitors (methanol and/or one of the glycols) are injected into the process where hydrates are likely to form. They push the hydrate formation equilibrium to lower temperatures or higher pressures. Certain inhibitors called low-dosage inhibitors keep any hydrate formed from coagulating and growing or delay the hydrate formation process.

4. Eliminate free water in the gas stream by dehydrating the gas or elevating the temperature to vaporize more water. Water is removed from the gas stream through a process called gas dehydration.
5. Redesigning piping systems (e.g., low points and restrictions).

Methanol, or one of the glycols, mixes with the condensed aqueous phase to lower its hydrate formation temperature at the given pressure (or increase the hydrate formation pressure at a given temperature) when injected into a gas process stream. Because they shift the equilibrium to lower temperatures and higher pressures, they are termed thermodynamic inhibitors. They also lower the freezing point of the liquid water (see Figure 2.14). To be effective, the chemical inhibitor is injected at the very points where the wet gas is cooled to its hydrate temperature. The injection must be in such a way that there is good mixing with the gas stream. Both the glycols and the methanol can be recovered together with the aqueous phase, then regenerated, and recycled. The choice between inhibitors depends on operating conditions and economics [3].

We look at hydrate inhibition methods and gas dehydration.

2.12.1.1 Hydrate inhibitors

Hydrate formation along a long natural gas pipeline has recently been established to initiate different types of internal corrosion along the pipe length based on the formation stage and point. These corrosions may lead to disintegration of the pipe's properties and eventually result in the pipe's leakage or full-bore rupture. Apart from the enormous economic implications on the operating organization, the conveyed fluid upon escape to the environment poses the risk of fire, reduction of air quality,, and other health hazards.

Hydrate inhibitors are used to inhibit hydrate formation, for example, methanol. Formation of hydrates may give rise to operational problems such as heat exchanger tube blocking, instrumentation plugging or pipeline blocking, and internal corrosion; hence, the conditions suitable for their formation be avoided whenever possible [3].

2.12.1.2 Methanol injection

Methanol has a low freezing point. It can be injected at any temperature and it is very effective as a hydrate inhibitor due to its ability to achieve high dew point suppression. Methanol is preferred for use at low temperature conditions when there is separation equipment downstream because glycol is harder to separate at low temperatures (due to its increased viscosity) [3].

Higher injection rates of methanol are required since some are lost to the hydrocarbon gas phase (due to its higher volatility) and some are lost to the liquid hydrocarbon phase (due to its solubility in liquid hydrocarbons). It is only the methanol that is dissolved in water that inhibits hydrate formation.

The efficiency of inhibition depends on the concentration of methanol injected. It can be injected intermittently or continuously [3].

2.12.1.2.1 Problems with methanol injection

Methanol can dissolve alcohol-based corrosion inhibitors injected into the process to prevent corrosion leading to unexpected corrosion problems.

Methanol can concentrate in the liquefied petroleum gas (LPG) stream. LPG is made largely of propane and mixed butanes. Methanol forms azeotropes with the propane and butane components in the LPG and these cannot be separated by distillation [3].

2.12.1.2.2 Methanol recovery

Recovering methanol from the process by distillation is not economical so in most instances, it is not recovered after use. In the gas plant methanol is not recovered.

2.12.1.3 Glycol injection

Ethylene glycol (EG) is the most used of all the glycols in hydrate inhibition due to its low cost, lower viscosity, and lower solubility in liquid hydrocarbons [3]. Losses are generally very small so do not need to be considered when calculating injection rates (see Figure 2.14).

2.12.1.3.1 Recovery and regeneration of glycol

It recovers and regenerates glycol. Following injection into the process, agitation or expansion from a high pressure to a lower pressure, like via a throttling valve, might cause the glycol-water solution and the liquid hydrocarbons to form an emulsion [3, 17, 20]. Glycol and the condensed water combination are extracted from the gas stream in a separator to recover the glycol [20]. After the water is removed from the recovered glycol-water mixture, the recovered glycol can be used again at the glycol regeneration unit [3, 17].

The regeneration process is designed to produce a glycol solution that must have a freezing point below the minimum temperature encountered in the system it is to be injected [3, 17, 20]. The freezing point of a glycol solution in water is dependent on the weight per cent of glycol in the solution. This is typically 75–80 wt% [3, 17, 20].

Glycol injection also aids in dehydrating the gas [3, 17, 20].

As noted earlier, the viscosity of Glycol increases as temperature decreases. The design of units containing chillers and refrigeration units where glycol is injected need to consider this. If the temperature is too low, the rich glycol solution leaving the unit is very viscous, making downstream separation difficult [3, 17, 20].

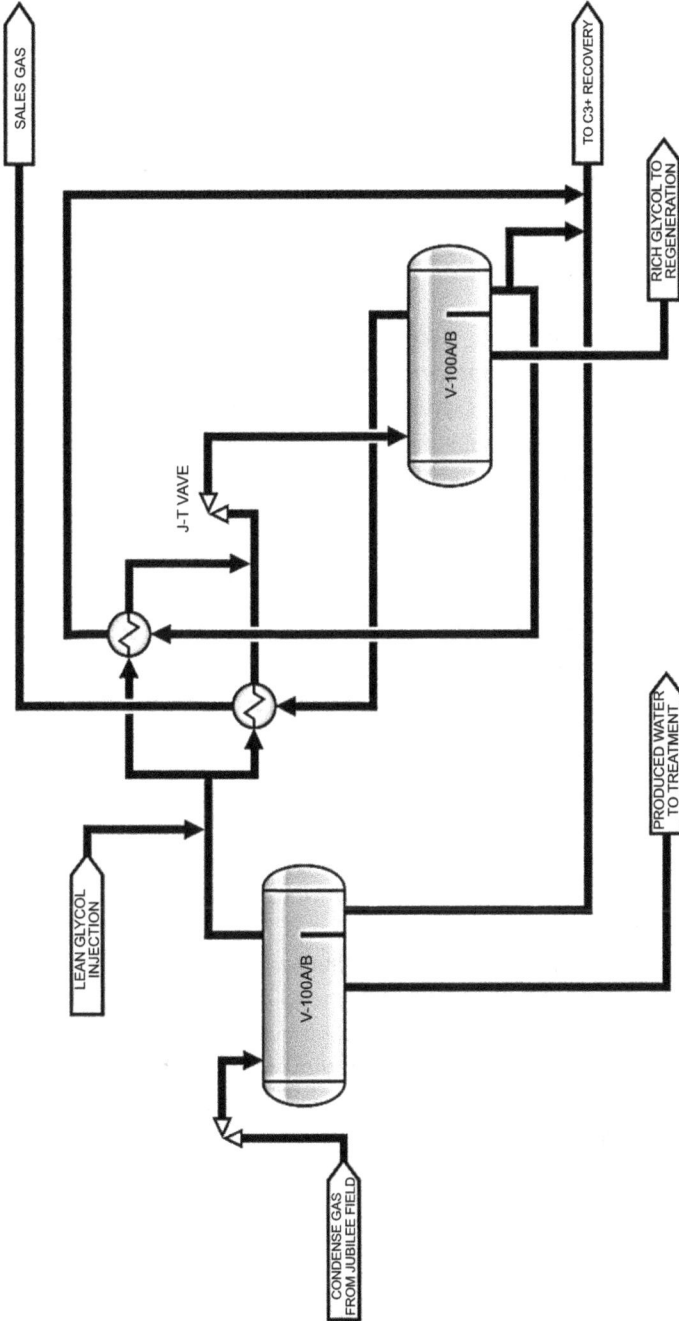

Figure 2.14: Glycol injection in the gas process plant [3].

2.12.1.4 Glycol dehydration

Avoid operational conditions that might cause the formation of hydrates by depressing the hydrate formation temperature using glycol dehydration. Refer to Section 2.5.3.4.1.2 for further information on glycol dehydration.

2.12.1.5 Water removal

If enough water can be removed from the produced fluids, hydrate formation should not occur. Dehydration is the common hydrate prevention technique applied to export pipelines.

For subsea production systems, subsea separation systems can reduce water flow in subsea flowlines. The advantage of applying subsea separation is not only hydrate control but also increasing recovery of reserves and/or accelerating recovery by making the produced fluid stream lighter and easier to lift. These systems need to be combined with another hydrate prevention technique such as continuous injection of a thermodynamic inhibitor or low-dosage hydrate inhibitor. The main risk associated with these subsea water separation systems is reliability [31].

2.12.1.6 Low-pressure operation

The pressure is maintained below the value necessary for hydrates to form. Decreasing the pressure below the pressure indicated on the hydrate formation curve for the given operating temperature [3].

2.12.1.7 Insulation

Insulation provides hydrate control by maintaining temperatures above hydrate formation conditions. Insulation extends the cool-down time before reaching hydrate formation temperatures [3].

When a pipeline carrying hydrocarbon fluid is shutdown for an extended period the temperature in the system should eventually come to equilibrium with the surroundings. Depending on the nature of the hydrocarbon being transported, remedial actions may be necessary during the period of shutdown before the temperature reaches the minimum allowable value. It is thus of interest to be able to predict, with reasonable accuracy, the length of time the fluid should take to cool down to any particular temperature. To ensure minimum cool-down time, the simplest solution is to insulate the pipe. Since the flow condition should change during the field's life, insulation and/or active heating should be adapted to the different conditions [3].

Insulation is normally not applied to gas production systems because the production fluid has low thermal mass and also should experience JT cooling. For gas systems, insulating is only applicable for high reservoir temperatures and/or short tieback lengths [31].

2.12.1.8 Active heating
Active heating includes electrical heating and hot fluid circulation heating in a bundle. In the flowlines and risers, active heating must be applied with thermal insulation to minimize power requirements [31].

2.12.2 Prevention of hydrate formation during commissioning

2.12.2.1 Pipelines
For pipelines that normally transport dry gas, the most likely occasion at which hydrate formation may pose a problem is during commissioning. Before commissioning with hydrocarbon gas, the pipeline should have been filled with water to carry out a system hydrotest. Hence, between the hydrotest and gassing-up operations free water must be removed or treated to prevent the formation of hydrates. Methanol (MeOH) or glycol should normally be used to dewater equipment that carries gas (pipelines, jumpers, etc.) to prevent hydrate formation during the bulk dewatering phase of the commissioning process [3].

Numerous methods of drying or treating the pipeline are
- methanol/glycol swabbing,
- vacuum-drying,
- hot air-drying, and
- inert gas-drying

For long pipelines, methanol swabbing offers the most economical solution. After a completed pipeline is filled with water and hydrostatically tested, conventional pigging runs remove the bulk of the water. Inevitably, there is a certain amount of water that passes the pigs and remains in the line. This residual water might form solid hydrates with the methane in natural gas under certain temperature and pressure conditions. Therefore, we use methanol or glycol swabbing, which involves passing a quantity of methanol down the pipeline between pigs [3].

This procedure leaves an aqueous film of methanol and water on the pipe wall that has a suitable concentration of methanol to inhibit hydrate formation when gassing up (any residual water present dissolves in the methanol, and any fluid passing the pigs contains only a small proportion of water, but a large proportion of methanol. This helps to prevent any subsequent formation of hydrates). From the volume of methanol used, the volume of liquid pigged out and the percentage of methanol in this liquid, estimates can be made of the volume of water still in the main, how long it should take to become dry and whether the methanol content of the aqueous film is high enough to prevent hydrate formation. Decisions can then be taken to either, gas the line up or pig again with methanol. After commissioning the concentration of methanol in the aqueous film must be maintained until the main is dry. This can be achieved by, either adding enough methanol to the gas to achieve this or adding more methanol at intervals to the line [3].

MEG is usually the fluid of choice for swabbing. MEG is typically chosen because it is easier to handle than methanol and can be regenerated.

The other drying procedures prevent the possibility of hydrates forming by completely vaporizing all free water in the pipeline.

2.12.2.2 Spool pieces

The final connection to the pipeline system usually comprises a double block valve and bleed system. During the precommissioning operations, both of these block valves should usually be closed, leaving the spool piece between the valves full of water [3] .

To prevent the possibility of hydrate formation during gassing-up this water must be treated. The simplest solution is to fill the spool piece with either methanol or glycol to a suitable concentration that should inhibit hydrate formation at the pipeline operating conditions. Ideally, as much water as possible be displaced, by the methanol or glycol, as this should further minimize corrosion problems that may occur if acid gases dissolve in the free water.

2.12.3 Removing hydrates

The existence of hydrocarbon hydrates in a system may be detected by a change in the pressure profile within the system. Hydrates, like any other obstacle in a pipeline, can be detected by the consequences they create. The following may indicate the presence of hydrates:

- reduced flow rates,
- reduced pressure,
- increased back pressure on a system,
- increased differential pressures, and
- temperature drops.

Once hydrocarbon hydrates have started to form the build-up accelerates. Hence, the system pressure profile should continue to change giving rise to potential capacity limitations.

The following methods can be used to remove hydrate plugs from a pipeline [3]:

1. Depressurization: The most common way to remove a hydrate plug from a flow channel is by depressurization. Flow is stopped, and the line is slowly depressurized from both ends of the plug. At atmospheric pressure, the hydrate stability temperature is invariably less than that of the surroundings, so heat flows from the environment into the hydrate plug. The plug melts radially inward, detaching first at the pipe wall.

2. Another method of removal of hydrocarbon hydrates from a pipeline system involves running a batch of either methanol or glycol through the pipeline, driven by a pig. The size of the batch should be sufficient to depress the hydrate formation temperatures of the maximum anticipated hydrate deposition, to below the pipeline temperature, as well as allowing for liquid that should be left on the pipe wall and not pushed forward by the pig. Methanol or glycol injection is usually ineffective because of the necessity of having the inhibitor contact the hydrate-plug face. When hydrates form in a vertical portion of a channel, such as a riser or well string, it may be possible to inject glycol or to place a heater at the plug face to promote hydrate dissociation.

3. Removal of the hydrates from raw natural gas pipelines, which may not have pigging facilities, is usually achieved by increasing the dosage rate of hydrate point depressant. For a short period, shock dosing, at a rate of approximately five times, the normal injection rate should be undertaken.

4. An alternative approach to removing hydrate is to operate the pipeline at different physical conditions. By lowering the operating pressure or increasing the operating temperature, the physical conditions suitable for hydrate formation may be avoided. This solution may be of limited use as the flexibility of the operating conditions, consistent with the required production rate, is unlikely to be high.

5. Coiled tubing represents the primary mechanical means of freeing the hydrate plug, but the maximum coiled-tubing distance is currently approximately 5 miles. Coiled tubing may be used to remove a substantial liquid hydrostatic head at the hydrate face to enable depressurization. Coiled tubing may also be used to inject methanol or glycol at the face of a hydrate plug, when density is insufficient to drive the inhibitor to the plug face.

6. Troubleshoot the reason for the hydrate or ice plug formation and remedy the problem, if possible.

2.13 Wax

2.13.1 Wax crystal modifier additives

Wax-related issues can arise in lubricating oils, distillates, residual fuels, and crude oil. The incorporation of additives into lubricating oils dates back to the 1930s. When the wax liquid cools below its cloud point, wax crystals begin to form and agglomerate. As the temperature continues to drop, these crystals accumulate to the extent that they create a loose gel structure. While this gel can be disrupted by shearing, it tends to reform when left undisturbed. The introduction of wax crystal modifiers can influence the growth and size of these crystals, thereby reducing their tendency to bond with one another. This process effectively lowers the temperature at which the gel structure forms during the cooling of the additive wax fluid, facilitating its movement through pipelines. Such substances are referred to as "freezing point depressants" [3, 32].

If the waxy oil is not treated, significant and costly handling problems can occur. These include
- loss of pipe capacity due to wax deposits on the inner walls of the pipe;
- development of no-flow conditions in the pipeline; and
- inability to restart flow after raw gels.

2.13.1.1 Uses
Wax crystal-modifying additives are considered a practical and cost-effective solution. Their application requires only modest capital and operating costs. Potential benefits include
- problem-free handling/use of waxy oils in pipelines;
- guaranteed restart ability; and
- nominal low-temperature pumping energy.

2.13.1.2 Applications
Wax crystal-modifying additives have demonstrated significant efficacy across various raw wax products, particularly in regions such as India, Indonesia, and the North Sea. A pertinent application can be observed with crude oil exhibiting a natural pour point around 30 °C, which is transported through a 200 km (125 miles) submarine pipeline. Given that the seabed's minimum temperature is approximately 20 °C, there is a risk that, in the event of a pump shutdown, the oil could cool to subfreezing temperatures.

To mitigate this risk, a wax crystal modifier additive produced by Lubrizol Corporation was utilized at predetermined treatment levels. At dosages ranging from 250 to 300 ppm, the pour point was effectively lowered to 12–15 °C, providing a substantial safety margin above the minimum seabed temperature of 20 °C.

Additionally, the wax crystal modifier branded as Shellswim, developed by Shell International Chemical Co. Ltd, has shown successful application in offshore oil pipeline systems, further underscoring the importance of these modifiers in maintaining flow assurance in challenging environmental conditions [3, 32].

2.13.2 Replacement for wax crystal modifier additives

While there are alternatives to using wax crystal modifiers as flow improvers, they are generally much costlier. Some of these alternatives include
- dilution of waxy oil with a lower wax content oil or other solvent options;
- heating wax oils beyond their crystallization temperature, necessitating the use of heaters or furnaces along with insulated piping or ducts to maintain the elevated temperatures. This setup incurs substantial operational costs.

2.14 Heavy and asphaltic crudes

Asphaltic and high-viscosity raw materials are generally unresponsive to wax crystal inhibitors or additives designed to reduce aerodynamic drag. However, Petroferm USA, a subsidiary of Petroleum Fermentations NV based in the Netherlands Antilles, has introduced an innovative polymer that facilitates the transport of heavy oil-water emulsions via standard unheated pipeline systems. This polymer not only aids in the handling of these emulsions but also permits their direct combustion, effectively allowing for the utilization of residual oil and water blends. This proprietary additive is referred to as Emulsan [3, 32].

2.14.1 Emulsan

Emulsan is a microbially derived biosurfactant that effectively stabilizes oil-in-water emulsions by inhibiting the coalescence of dispersed oil droplets and preventing the reversion to a more viscous phase. It is commonly used in conjunction with conventional surfactants, which rapidly migrate to the oil-water interface and facilitate initial emulsion formation. Due to its strong affinity for the oil-water interface, Emulsan may influence the microbial degradation pathways of emulsified hydrocarbons, which has significant implications for the emulsion's stability during both storage and transportation as well as its environmental biodegradability in the event of a spill. A key benefit of Emulsan lies in its ability to retard coalescence, helping maintain emulsion integrity while it reverts to its original oil-water separation state [2, 34].

Emulsan has potential uses in the petroleum industry, including the formation of heavy oil-water emulsions to reduce viscosity during pipeline transport and the production of fuel oil-water emulsions for direct combustion with dewatering [3, 32].

2.15 Paraffin

Heavy paraffins consist of normal hydrocarbons ranging from approximately $C_{18}H_{38}$ to $C_{38}H_{78}$, typically intermixed with smaller quantities of branched paraffins, monocyclic paraffins, polycyclic paraffins, and aromatic compounds. The concentration of these paraffins in petroleum can fluctuate widely, ranging from less than 1% to over 30%.

Studies indicate that the characteristics of paraffin deposits – such as quantity, hardness, adhesion, wax content, and average molecular weight – are primarily influenced by surface roughness, assuming other conditions remain constant. Paraffin deposition is a recognized challenge in oil wells, leading to significant operational issues.

To mitigate these problems, internal coatings have been employed to minimize environmental impact. Evaluations of various plastic coatings reveal that while most smooth, nonparaffinic plastics can be effective, flexible, highly polar, nonparaffinic plastics demonstrate superior long-term resistance to paraffin deposition, particularly in environments where the flowing stream contains abrasive particles [13].

In situ, plastic coating of oil wells is a specialized method employed to safeguard the internal surfaces of oil well pipes and tubing from corrosion, wear, and scaling. Various methods exist for the removal of accumulated paraffin waxes in oil wells, including combined treatment approaches, mechanical cleaning, scratch techniques, coiled tubing applications, as well as thermal methods such as heating, hot oiling, and hot water flushing. Additionally, chemical treatments utilizing wax solvents and dispersants are also utilized to effectively mitigate wax buildup.

3 Risk assessment

ISO 31000 Risk Management suggestions provide a clear definition of risk because they are manufactured from two number one elements: the chance of failure (POF) and the ensuing outcomes of the failure (COF). This definition is extensively generic and forms the premise of risk control practices in various industries. For the duration of the life cycle of a pipeline or pipeline machine, beginning from construction and continuing through decommissioning and abandonment, diverse dangerous elements are present. These dangers can pose threats to lifestyles and health, endanger the surroundings and property, and create possibilities for financial loss. This segment pursues to clarify the danger assessments associated with each pipeline and pipeline structure at the same time as outlining a way to identify dangers within the pipeline system [1, 3]. Chance evaluation performs an essential function in integrity control. It encompasses both the danger analysis and hazard evaluation stages of the pipeline gadget. The procedure of hazard assessment needs to begin with a conceptual layout and be evaluated regarding operation, preservation, tracking, modifications, checking out, and abandonment [1]. As a part of an integrity control approach, risk exams need to be performed on pipeline structures and their related centers, together with meter and valve stations. These checks might also observe both a prescriptive-based or overall performance-based method [1]. If a hazard evaluation isn't accomplished, the operators of the pipeline device have to file their reasons and prioritize all upcoming pipeline integrity evaluations as well as preventative and mitigation measures while setting intervals for integrity checks. Prescriptive-based programs utilize hazard exams to persuade selections and prioritize planned sports. In evaluation, performance-based total packages use chance tests to help in planning and prioritizing sports, determining essential inspection, prevention, and/or mitigation obligations, and organizing related statistics and information [1, 3].

3.1 Objectives

The risk assessment empowers the pipeline system operator to confidently determine the following:
- rank of precedence sections for inspection;
- energetic threats for which integrity tests are wished; and
- mitigation measures to put in the area primarily based on the impact on danger reduction.

Risk exams are a crucial device in maintaining the integrity of a machine by evaluating both the potential impact and likelihood of incidents. This method encourages rational and steady decision-making, bearing in mind the identity of vulnerable areas

https://doi.org/10.1515/9783111629742-003

and the implementation of suitable mitigation measures. Through analyzing both primary threat factors, such as POF and COF, the method avoids the pitfall of entirely addressing visible or frequently taking place troubles at the same time as neglecting probably catastrophic events. At the same time, it prevents the oversight of extra likely scenarios by focusing entirely on much less probably however intense events [1].

3.2 Risk evaluation technique

The threat analysis technique is a crucial factor of the integrity control application, employed to assess the potential influences of risks on a pipeline or pipeline gadget [1]. Through carrying out an intensive danger analysis, it's far viable to pick out issues which could arise for the duration of precise phases of the pipeline or pipeline machine's life cycle [1, 2, 21]. A danger observation (evaluation) ought to consist of the subsequent elements [1]:

- definition of the objectives of the evaluation
- justification for carrying out the evaluation
- identification of described damaging consequences and issues
- will power of appropriate chance size technique
- description of the pipeline system, which includes
  - assessment of the pipeline, encompassing its reason, capability, and geographic location
  - specifications of dimensions and substances used in the pipeline, inclusive of coatings and ancillary equipment [1]
  - evaluation of the condition of the pipeline, coatings, and ancillary components
  - examination of operational conditions, which includes pressures, temperatures, and the character of carrier fluids
  - description of the physical and geographical surroundings surrounding the pipeline
  - delineation of the bodily boundaries relevant to the threat analysis [1]

3.3 Risk estimation

The manner of assessing the findings obtained from the chance evaluation, together with the analytical critiques carried out, is referred to as hazard assessment.

This process not handiest includes a scientific review of the diagnosed dangers but additionally entails a radical evaluation of their ability to affect the organization's objectives [1].

Moreover, hazard assessment encompasses the identity and assessment of diverse options to be had for efficiently coping with those risks.

This includes analyzing one-of-a-kind danger remedy techniques and thinking about their potential effects, in addition to their related charges and advantages, to decide the most suitable approach. By undertaking this comprehensive assessment, businesses could make knowledgeable decisions that align with their threat tolerance and beautify typical resilience [1].

3.4 Risk evaluation

The process of assessing the findings obtained from the risk assessment, along with the analytical evaluations conducted, is referred to as risk evaluation. This process not only includes a systematic review of the identified risks but also entails a thorough evaluation of their potential impact on the organization's objectives [1].

Furthermore, risk evaluation encompasses the identification and assessment of various options available for effectively managing these risks. This involves analyzing different risk treatment strategies, considering their potential consequences, as well as their associated costs and benefits, to determine the most suitable approach. By engaging in this comprehensive evaluation, organizations can make informed decisions that align with their risk tolerance and enhance overall resilience [1].

3.5 Risk significance

Risk significance assessments entail the evaluation of the importance of estimated risk levels to individuals impacted during an incident or hazardous event as well as the identification of those responsible for safeguarding the interests of all parties involved in the incident [1, 3, 26]. The significance attributed to the estimated risk is contingent upon the context in which the risk assessment is conducted. The following factors should be considered in determining risk significance [1]:
- The potential severity of the consequences resulting from the incident [1]
- The anticipated frequency or likelihood of the incident's occurrence
- The societal benefits provided by the risk source to both the community and those potentially affected by the incident [1]
- The costs associated with mitigating the estimated risk level

Guidance for determining risk significance can be derived from the application of the following elements [1]:
- Comparing the estimated risk level against other recognized activities and events
- Reviewing national and international precedents established in various industries as well as consulting the literature on risk acceptance criteria [1]
- Drawing on in-house experience and existing guidelines to effectively identify hazardous situations

Should a significant level of risk be identified, the following response measures are required:
- Conducting a more detailed risk analysis to reduce the uncertainties or ambiguities associated with the estimations that may have resulted in an overestimation of risk [1]
- Considering available options to diminish the estimated risk level [1]

3.5.1 Options analysis

Following the completion of the risk evaluation phase, the next critical step involves the identification and thorough analysis of various options that can effectively reduce risks. This process is particularly important when the risk significance has been assessed as high, indicating a pressing need for intervention [1]. The goal of this analysis is to develop risk reduction strategies that are not only proactive but also tailored to address both the likelihood of occurrence and the potential severity of consequences associated with hazardous events [1].

Risk reduction strategies may include a diverse range of measures such as engineering controls, safety protocols, training programs, and contingency plans. These strategies aim to mitigate the frequency of incidents and lessen the impact on individuals, property, and the environment should an adverse event occur [1].

A crucial element of this analysis involves evaluating the risk estimate within the context of a base-case scenario. This serves as a benchmark against which various alternative scenarios can be compared. Through a comprehensive option analysis, different hypothetical situations are explored, allowing decision-makers to visualize potential outcomes and the effectiveness of various risk management approaches [1].

This analytical framework is foundational to the decision-making process, guiding stakeholders in determining the most appropriate courses of action for managing risks associated with identified hazardous events. The insights garnered from this rigorous evaluation will aid in prioritizing interventions and resource allocation, ensuring that efforts are aimed at the most significant risks present in the environment [1, 26].

3.6 Documentation

The risk assessment process needs to be documented in a risk assessment report. This report should consider the goals of the assessment and the individuals who will be reading and evaluating it:
- The report should discuss both the benefits and drawbacks of the different methods used to reduce risks as well as any uncertainties related to the risk estimates generated during the analysis [1].

- The length and detail of the risk assessment report will depend on its goals and the specific situation being assessed. However, it must include the following key sections [1]:
- Objectives and scope: What the assessment aims to achieve and its boundaries [1].
- System description: A breakdown of the system or situation being assessed [1].
- Risk analysis methodology: The approach taken to analyze the risks.
- Limitations and assumptions: Any constraints and basic assumptions made during the assessment [1].
- Hazard identification results: The dangers or risks identified [1].
- Frequency analysis and assumptions: How often these risks might occur and the assumptions behind those estimates [1].
- Consequential analysis and assumptions: The potential outcomes of the identified risks and the assumptions made about them [1].
- Risk estimation results: The estimated level of risk based on the analysis.
- Sensitivity and uncertainty analysis: An examination of how sensitive the results are to changes and the uncertainties involved [1].
- Discussion minutes: A record of discussions about the results including any problems identified during analysis [1].
- Conclusions and recommendations: Final thoughts and suggestions based on the assessment [1].
- References: A list of all sources that support the methods and techniques used.
- Names and qualifications: The names and qualifications of the people who were involved in the assessment process [1].
- This structured approach helps ensure that everyone involved can understand the findings and make informed decisions based on the assessment [1, 26].

4 Investigation

In the event of an incident, such as a leak or rupture within a pipeline or its auxiliary components, it is imperative to initiate a pipeline incident investigation [1]. This investigation aims to identify the underlying cause of the failure and to implement preventive measures to avert the recurrence of similar incidents, as outlined in LI 2189, Articles 11.34 and 11.35, as well as in CSA Z662, Article 10.2.3 (latest edition) or IGE/TD/1 [8, 25]:

- In instances where there is a high frequency of failures associated with a specific pipe or a group of pipes, it is imperative to implement appropriate mitigating measures or consider the replacement of the affected piping systems [1].
- If an investigation fails to establish a definitive cause for the failures, it may be prudent to retest the pipeline and contemplate a reduction in operating pressure as a potential strategy to diminish the likelihood of future failures [1].

4.1 Incident investigation process

In the event of a pipeline failure, it is essential to implement specific elements as part of a comprehensive incident investigation. This approach facilitates the acquisition of valuable information from the incident, which can subsequently inform the development of procedures and practices aimed at preventing future occurrences:

- Incident detection and product release [1]
  - Report the incident to a supervisor promptly.
  - A manager or designated representative will subsequently arrive at the location of the incident.
- Collection of necessary data as follows:
  - Release location (mapping, GPS, and physical topography of the area)
  - Size and impact of the release using a grid drawing
  - Pipeline and product data, weather, and timeline of events
- Documentation information
  - Photographs of the pre-excavation area and the landscape around the failure site
  - Initial photographs of excavations, fault area, and soil types
  - Initial suspicion of the cause of the failure (i.e., internal/external corrosion and third-party contact)
  - Operational data at the time of the incident (pressures, product temperature, weather, and priority level)
  - Photographs and forms of exposed piping and cutout area
  - Photographs and inspection form of pipe coating

https://doi.org/10.1515/9783111629742-004

- Sampling
  - Contaminated and uncontaminated soil around the site
  - Groundwater, contaminated, and uncontaminated water or product primer or coating (if any)
  - Any solids primer or internal pipe cutout (if any)
  - Sample of pipe corrosion (if applicable)
- Sampling of the cut pipe
  - Do not touch the surface of the pipe or the crack. Leave it as it was found.
  - Protect all found pipe fragments with protective wrapping.
  - Record pipe information such as line number, outside diameter, inside diameter, and grade on the sample and keep it with the cut or fragment.
  - Complete the cutout form if available.

5 Inspection, monitoring, and repair

The pipeline integrity engineer and a corrosion technician maintain the corrosion control program. External corrosion control of buried facilities is primarily achieved through the use of coatings supported by cathodic protection (CP).

The pipeline integrity engineer is responsible for operating and maintaining CP equipment. The pipeline integrity engineer should be responsible for the monthly rectifier and the pipe-to-soil readings.

The external protection of the pipeline and ancillary facilities is achieved by a system comprising coating and CP. The performance of this system is assessed by regular monitoring of pipe-to-soil potentials at selected intervals along the pipeline. From these results, conclusions may be drawn concerning the level of CP being achieved and the performance of the coating [1, 3].

A comprehensive monitoring and maintenance plan is required to be developed, based on the results of the risk assessment and integrity assessment, as well as the applicable regulations. The plan should encompass the following items: monitoring of the pipeline route, pipeline maintenance, and maintenance of pipeline stations. The plan should clearly define the monitoring and maintenance activities for the areas of concern identified through the risk assessment or integrity assessment. All maintenance and testing of protection equipment should adhere to documented procedures. The plan should be reviewed periodically per the update of the risk assessment or integrity assessment, and performance review.

Monitoring activities include the identification and prevention of external threats to the pipeline, the assessment of the effectiveness of the CP to prevent external corrosion, and the detection of any changes in the environment that could increase the risks. The prevention activities aim to control any external activities on the pipeline and prevent external corrosion, the main threats to which a transmission pipeline is subjected. These activities include pipeline patrol, management of third-party activities, monitoring of the effectiveness of the protection systems against external corrosion, maintenance of the right of way (ROW) of underground pipelines, maintenance of aboveground sections, maintenance of the CP equipment, and maintenance inspections at specific points [3].

The purpose of the maintenance inspection of aboveground sections is to verify the visual inspection of all aboveground piping sections for damage, the condition of the painting, the proper condition of mechanical protection equipment (including fences of stations, accesses, pipe supports), condition of the antifriction pads between the pipe and the steel part of the support, good state of marking, identification, and other means of information to the public. If the aboveground section is isolated from the buried pipeline by insulating joints, check adequate grounding and electrical continuity of piping and supporting structures. If the aboveground section is electrically connected to the buried pipeline, check proper isolation at all supports and proper condi-

https://doi.org/10.1515/9783111629742-005

tion of protections that prevent unauthorized access. The risers of the aboveground section should be checked for corrosion, and the coating should be removed and replaced if found to be defective (e.g., cracks or blistering) [3].

Remember these important steps when getting ready for an in-line inspection (ILI) tool run:

1. Identify threats: First, understand the potential problems with the pipeline. Talk to your pipeline integrity engineer or subject matter expert (SME) to figure out what issues to look for. This will help you select the right ILI tools.
2. Prioritize inspections: Decide which parts of the pipeline need to be checked first based on how often they need reinspection and how risky they are. You should at least use tools to check for bending and rust.
3. Choose a vendor: It can be hard to pick a good ILI vendor because there are so many. Look at vendors closely to make sure they can do what you need.
4. Learn from the past: See what you can learn from previous ILI runs. For example, if the tool has lots of starts and stops, look into it. Change how fast the liquid moves and consider changes in pipe thickness to prevent problems.
5. Include ECA analysis: Along with making physical changes to fix the pipeline, think about doing engineering critical assessments (ECAs). ECAs help you spot and deal with threats and make sure the pipeline follows the rules.

Remember that working with experts and learning from what's happened before are what make ILI preparations go well.

5.1 Inspection

Pipeline inspection constitutes a vital element in preserving the integrity of pipeline systems and generating data necessary for evaluating their current and future conditions. Consequently, it is essential for inspection strategies to delineate the required inspections, their scope, and the appropriate intervals for conducting such inspections. A risk-based inspection methodology should be implemented to ensure that the techniques employed, the extent of coverage, and the intervals are commensurate with the associated risk levels. The integrity of a pipeline may be compromised by various mechanisms or threats, including:

- internal corrosion,
- external corrosion,
- third-party damage,
- fatigue,
- mechanical overload,
- crack-like flaws, and
- manufacturing defects.

Different inspection techniques must be employed based on the specific location of the pipeline section and the identified threats. It is essential for inspection strategies to delineate the requirements for inspection tools and techniques corresponding to each failure mechanism or threat. The scope of inspection coverage and intervals may be determined through risk assessments, time-based scheduling, or ad hoc evaluations. Within the framework of the pipeline integrity management system (PIMS), it is imperative to specify the types of inspections to be conducted, the coverage parameters (e.g., ensuring 100% examination of the pipeline length across its full circumference), the frequency of inspections, and the rationale for the established inspection intervals. At a minimum, the following inspection activities shall be executed to effectively identify threats within pipelines [3].

Patrolling the pipeline constitutes a vital visual inspection activity that enables the assessment of surface conditions, identification of leak indications, observation of construction activities, and evaluation of other factors that may influence the safety, integrity, and security of the pipeline system. This patrolling is conducted along onshore pipelines and laterals utilizing appropriate methods, which may include walking, driving, flying, or other suitable means of inspecting the ROW. During the patrolling process, our personnel exercise due diligence and compile a list of observations that would be categorized as "reportable observations" should they be encountered while traversing the ROW.

For aboveground pipelines, laterals, and block valve stations, we mandate a closed visual examination coupled with an ultrasonic wall thickness inspection survey to establish the absence of significant external or internal corrosion, as well as to detect other anomalies such as mechanical or coating damage. A pit depth gauge or other appropriate instruments are employed to measure the depth of any identified external corrosion. Additionally, ultrasonic tools are utilized to ascertain the wall thickness of the pipeline and to quantify the degree of internal corrosion. Particular attention is devoted to external corrosion found beneath clamps, supports, sleeves, and insulating materials. Our visual inspection of onshore pipelines also encompasses an evaluation of the condition of the coating, including paint peeling, blistering, or deposits present on the pipe surface, as well as the condition of pipe supports and any indications of mechanical damage affecting both the pipeline and its supports.

We conduct ILIs, commonly referred to as intelligent pigging, of pipelines utilizing magnetic flux leakage (MFL) and/or ultrasonic testing (UT) technologies. Our preference is to employ the UT tool when the pig driving medium is water. Additionally, we consider incorporating an extra tool to acquire coordinate mapping and geometric details of the pipelines, including bends, dents, ovalities, and other features, alongside the metal loss assessment tool (MFL/UT). The frequency of ILIs shall be determined based on risk assessment or established time intervals; however, such inspections shall not exceed half of the pipeline's remaining lifespan, with a maximum inspection interval of 10 years. We are confident that this methodology will enhance the integrity, safety, and security of our pipeline systems.

To evaluate the protection level and external condition of buried pipelines and their protective coatings, it is imperative to conduct CP ON-OFF, close interval potential surveys (CIPSs), and direct current voltage gradient (DCVG) surveys. Should any anomalies be identified, excavation may be warranted to assess the pipeline's condition and ascertain the cause of any localized damage. CIPS and DCVG surveys should adhere to either a risk-based or time-based schedule, with a maximum frequency of every 5 years, while ON-OFF surveys should be executed on an annual basis. CP monitoring facilities must be installed along the pipeline route to facilitate the assessment of CP performance, in accordance with ISO 15589-1 [29]. Pipeline owners are encouraged to consider extending the intervals for CP inspections if the pipelines are subject to direct current stray current interference. An evaluation of the stray current corrosion risk associated with the pipeline should be conducted to determine appropriate inspection frequencies. In cases where a pipeline is exposed to significant levels of direct current stray current interference, the pipeline owner should re-evaluate the ILI intervals and correspondingly increase the frequency of such inspections [2].

Pipeline emergency shutdown valves (ESDVs) are critical safety equipment for pipeline inventory isolation. Periodic testing of ESDVs is necessary to ensure they function on demand and comply with safety design requirements for valve closure time. Seal leak testing should also be performed to ensure that the closed valve is secure and does not allow fluid to pass through. CP ON-OFF, CIPS, and DCVG surveys shall also be performed to evaluate the level of protection, the external condition of the buried pipelines, and their protective coatings. If an anomaly is detected, excavation may be required to physically check the external condition of the pipeline and understand the reason for the localized damages. CIPS and DCVG survey frequencies shall be either risk-based or time-based but shall not exceed 5 years. ON-OFF survey shall be carried out yearly. CP monitoring facilities along the pipeline route to allow measurement of the performance of the CP shall be installed per ISO 15589-1 [33]. Pipeline owners shall consider more frequent CP inspection intervals if pipelines are exposed to DC stray current interference. An assessment of the level of stray current corrosion risk on the pipeline shall be carried out to determine the inspection frequency. If a pipeline is exposed to significant levels of DC stray current interference, then the pipeline owner shall review the ILI interval as well and increase the frequency at which ILI is carried out.

Pipeline ESDVs are considered safety-critical equipment for the isolation of the pipeline inventory or part of it. ESDVs shall be periodically tested to ensure that they will be functioning on demand and that valve closure time is compliant with the safety design requirements. Seal leak testing shall also be carried out to ensure that the closed valve is holding tight and does not allow the fluid to pass through [8].

5.1.1 Internal inspection

Pipelines must undergo internal inspection using ILI tools. It is essential to conduct a series of pigging runs, encompassing cleaning, geometric, and profiling pigs, to facilitate the safe passage of the inspection pig. The implementation of an MFL pig or an ultrasonic pig is crucial for identifying metal loss in the pipe walls, effectively locating and sizing areas affected by corrosion and mechanical damage. Additionally, the MFL pig is adept at detecting circumferential cracks.

For pipelines that are prone to stress corrosion cracking, it is advisable to consider employing elastic wave or transverse flux leakage internal inspection tools to enhance detection capabilities in these critical areas.

5.1.2 External inspection

If the use of an internal ILI tool is not feasible, then an external aboveground survey should be undertaken. The following external survey methods may be used:
1. CIPS
   The CIPS survey determines the actual level of CP being experienced along the pipeline by measuring the pipe to soil potential and hence corrosion protection levels, and quality of coating.
2. DCVG survey
   The DCVG survey should be used for a detailed assessment of the condition of the pipeline coating. DCVG should be used for coating defect size evaluation, defect length evaluation, defect corrosion status, and defect influence regarding electrical interference [3].
3. Pearson survey
   The Pearson survey should also be used to locate coating defects. It is very effective in tracing discontinuities or damage in buried pipelines' coating, as well as loose electrical contacts and their exact site on the pipe, allowing the prevention of major failures [3].

5.2 Mandatory monitoring

Mandatory inspections and monitoring that should be conducted regularly are:
1. CP surveys – It is desirable to take a series of electrical measurements on a newly installed buried CP system to determine the initial level of CP. After the initial series, measurements should be made after 6 months and 1 year of operation. This should enable the corrosion engineer to identify deficiencies and program corrective action. After the first year of operation, measurements should be made

at least annually, unless the condition indicates more frequent testing (refer to Section 192.465 of 49 CFR Part 192 Subpart I).

The effectiveness of CP systems should be determined by comparing survey results to criteria listed in Clause 6.2 of NACE SP0169-2013 or Clause 5.3 of ISO 15589-1. Any deficiencies or repairs that are outlined in the report and compromise the effectiveness of the CP system should be addressed within 12 months of discovery.

2. Rectifier monitoring/maintenance – All rectifiers and critical current sources should be inspected at least once every 2 months for proper operation. Monthly inspections of CP rectifiers must be made to make sure that the units are operating as intended for their specific purpose. The inspections should include measurements of current and voltage output and a check of components for damage or deterioration.

3. ROW patrolling/inspection – Annual ROW inspections should be conducted on pipelines with water crossings or unstable ground, or on any pipeline identified as a high-risk pipeline. The frequency of ROW inspections is dependent upon operating conditions as per the pipeline operations and maintenance manual. Any issues or items that pose a threat to pipeline integrity should be addressed within 6 months of inspection.

4. Annual internal corrosion review – Once per calendar year, the internal corrosion susceptibility of each licensed pipeline should be reviewed by assessing one or all of the following data types:
   - Production and operating parameters
   - Gas analysis
   - Corrosion monitoring data
   - Mitigation program records
   - Inspection and repair records

5.2.1 Supplemental integrity monitoring

1. Corrosion monitoring devices – Pipelines that display proven internal corrosion activity, or are deemed as high-risk to internal corrosion activity, should be monitored via corrosion coupons, probes, nondestructive testing, or other means of assessing pipe wall condition. Data from newly installed monitoring devices should be collected at intervals not exceeding 6 months until a new collection frequency is determined by historical corrosion rates.

2. Water crossing inspections – At water crossing locations where possible, soil erosion is identified, scour, or slope instability, additional surveys such as the depth of cover and/or hydrotechnical evaluations should occur within 6 months of the discovery of the threat.

5.2.2 General

Standards for all construction projects that require new CP installations are detailed and designed by a corrosion consultant.

Once each calendar year, the pipeline system should be surveyed to ensure that adequate CP is being applied to all areas of the system. The services of a corrosion consultant can be used to perform these annual surveys and to recommend any remedial repairs or additions to the CP systems. The pipeline integrity engineer, along with the corrosion consultant, should then analyze these reports and produce a program of repairs, replacements, and new installations based on the analysis report. The annual potential survey consists of readings taken at every test lead, mainline valve, piping elbow, tank, or any other equipment attached to the system that requires CP applied to it. Readings are also taken at all casings, foreign crossings, and insulating fittings between the pipeline and external operator's tie-in points.

5.2.2.1 Guideline

Adjustive surveys of CP systems should be performed annually and include the following:

– Complete inspection of the rectifier and ground bed installation to determine overall condition and efficiency with corrective measures taken as required.
– Complete inspection of all resident pipe-to-soil meters to determine condition and accuracy. Repair and recalibrate as required.
– Rectifier output levels adjusted to predetermined levels and pipe-to-soil potentials measured at provided pipeline test leads and resident pipe-to-soil meters. Faulty test leads discovered during the survey are repaired as required. Rectifier output levels are readjusted during a pipe-to-soil potential survey to ensure pipe-to-soil potentials are maintained at adequate and balanced protective levels on all pipelines and in their entirety.

All cased road crossings are inspected and measurements taken to determine if contact between carrier pipe and casing is evident, and, where there is evidence of contact, corrective action is taken as required [3]:

– Evidence of foreign interference is investigated and corrective measures implemented.
– Inspection and repair or replacement of faulty insulating material or fittings as determined from pipe-to-soil potential measurements.
– Completed and detailed records of all adjustive survey data (repairs and maintenance, measurements, settings, adjustments, etc.) are compiled and recorded for reference.

When the pipe to soil potential survey indicates inadequate protection, check at the rectifier unit. The current output of the rectifier should be adjusted:

- If the current is high and accompanied by low voltage, suspect the pipeline and investigate for a possible short-circuit to another metal structure.
- If the current output is low, with normal or high voltage, suspect the anode ground bed or connecting cables.

5.3 Repair

This section details the appropriate response protocols for findings identified during inspections. Actions can be categorized into two primary types: corrective repairs aimed at eliminating unsafe conditions, and preventative measures designed to mitigate or eliminate potential threats via maintenance. The frequency of inspections will be dictated by the nature of the identified defect, the efficacy of the implemented mitigation strategies, and the preventive measures in place.

Response actions are classified into three distinct categories: immediate response, scheduled response, and monitored response. An immediate response is warranted for defects at impending failure thresholds. Scheduled responses are required for defects deemed significant but not critical enough to necessitate immediate action; these can be addressed prior to the next scheduled inspection. A monitored response applies to defects that are not anticipated to reach a failure state before the subsequent inspection.

To ensure the integrity of the pipeline system, it is crucial to select and time the appropriate examination, evaluation, and mitigation actions within the integrity management program. This program must also include a rigorous analysis of both existing and newly implemented mitigation actions to assess their effectiveness and validate their continued application moving forward.

5.3.1 Repair objective

Repair procedures are crucial for safeguarding the well-being of workers, employees, and the public, while also mitigating environmental risks. Establishing clear protocols for acceptable work practices is essential for achieving these objectives.

The structural integrity of the pipeline is fundamental to its safe operation. To uphold this integrity, any identified anomalies, whether external or internal, must undergo thorough assessment and, if warranted, repair. The primary goal of the repair process is to restore the pipeline to its original state of integrity or to ensure that any remaining anomalies are managed so that the pipeline can continue to operate safely [3].

Pipeline repairs may be necessitated by a variety of factors, including but not limited to:
- Damage from third parties
- Manufacturing defects identified during the rolling phase
- Damage sustained during pipe handling and installation processes
- Internal and external corrosion
- Structural integrity issues such as cracks and dents
- Deformation like buckles
- Surface imperfections including gouges

The approach to pipeline repair is contingent upon the severity of the damage and the cost-effectiveness of the repair strategy. Various methods exist for executing repairs, tailored to the specific conditions and requirements presented by the pipeline's state.

5.3.2 Pipeline repair standard

When completing repairs on a pipeline, current codes, practices, and regulatory policies must be adhered to. This section provides guidance on defect evaluation and describes the repair methods for the different types of pipeline anomalies. These standards should be complied with when evaluating an anomaly for potential repair [1, 3]. These standards are:
- LI 2189: Natural Gas Pipeline Safety (Construction, Operation and Maintenance) Regulations, 2012
- CSA Z662 (Latest Edition): Oil and Gas Pipeline Systems
- ASME B31G-1991 R2004: Manual for Determining the Remaining Strength of Corroded Pipelines. (This is a supplement to ASME B31 code for pressure piping.)
- API 1104: Standard for Welding Pipelines and Related Facilities
- ANSI/API RP 579: Fitness-for-Service (Recommended Practice)
- DNV-RP-F113: Pipeline Subsea Repair – Rules and Standards
- DNV-RP-F102: Pipeline Field Joint Coating and Field Repair of Line Pipe Coating
- ASME B31.8S: Managing System Integrity of Gas Pipelines

5.3.3 Corrosion

Corrosion can be generalized as a loss of pipe wall thickness and can be categorized as general surface corrosion, localized external pitting corrosion, localized internal pitting corrosion, or some combination of the above. When evaluating a corrosion anomaly, both of the above standards are to be considered.

Before evaluating the corrosion anomaly, the corroded areas should be thoroughly cleaned, to remove corrosion products, so that their dimensions can be measured accurately. For anomalies with a maximum depth of 10% or less of the nominal pipe wall thickness, no repair is required.

For anomalies with a depth greater than 10% and less than 80% of the nominal wall thickness of the pipe, an evaluation is required to determine whether a repair is required. The anomaly should be evaluated by the criteria in ASME B31.8S or Clause 14.46 of LI 2189 [8, 24]. These criteria should calculate the maximum allowable longitudinal length of the corroded area. [ASME B31G-1991 R2004 contains the identical criteria to CSA Z662 (latest edition) for determining the maximum allowable longitudinal length of the corroded area] [21].

If the corroded area is greater than the maximum allowable longitudinal length, a safe maximum operating pressure (P) can be calculated using the ASME B31G-1991 R2004 standard. According to ASME B31G-1991 R2004, if the established MAOP of the pipeline is equal to or less than P, the corroded area may still be used for service at that MAOP. If the MAOP is greater than P, then either
– a lower MAOP should be established such that it does not exceed P, or
– the corroded area must be repaired or replaced.

In order to ensure the safety of the pipeline system, immediate response is crucial for any corroded areas that have a predicted failure pressure level less than 1.1 times the maximum allowable operating pressure (MAOP) as determined by ASME B31G or equivalent. It is important to note that any metal-loss indication affecting a detected longitudinal seam also falls under this category. To confirm these indications after determining their condition, the pipeline system operator must act promptly. In case a defect requiring repair or removal is found after examination and evaluation, it should be remediated promptly unless the operating pressure can be lowered to mitigate the need for repair or removal.

Scheduled responses are appropriate for certain indications that allow for continued operation. Indications with a predicted failure pressure greater than 1.1 times the MAOP should be confirmed and evaluated. It is pertinent to prioritize these indications and take necessary actions to ensure the safety of the pipeline system.

Monitored responses are the least severe indications and do not require confirmation and evaluation until the next scheduled integrity assessment interval stipulated by the integrity management plan, provided they are not expected to reach critical dimensions before the next assessment. It is crucial to keep a close check on these indications to ensure they do not escalate to severe levels. In this case, the indications requiring immediate response due to immediate or near-term leaks or ruptures would be any corroded areas that have a predicted failure pressure level less than 1.1 times the MAOP as determined by ASME B31G or equivalent. Also, in this group there would be any metal-loss indication affecting a detected longitudinal seam. The pipeline system operator shall promptly confirm these indications following determination of the condition. After examination and evaluation, any defect found to require repair or removal shall be promptly remediated by repair or removal unless the operating pressure is lowered to mitigate the need to repair or remove the defect. Scheduled responses to certain indications are suitable for continued operation. Indications characterized with a predicted failure pressure greater than 1.1 times the MAOP shall be confirmed and evaluated. Monitored responses are the least severe indications and do not require confirmation and evaluation until the next scheduled integrity assessment interval stipulated by the integrity management plan, provided that they are not expected to grow to critical dimensions prior to the next scheduled assessment.

5.3.3.1 Gouges, grooves, and arc burns

All types of gouges, grooves, and arc burns should be considered to be a defect as per ASME B31.8S or Clauses 6.2–6.7 of LI 2189, gouges, grooves, and arc burns. If a defect of this type is located within the pipe, it should be removed by grinding until the ground area blends smoothly into the adjacent pipe. All repairs must be completed in accordance with a standard such as ASME B31.8S or Clauses 6.2–6.7 of LI 2189 (latest edition), grinding repairs [8, 24].

5.3.3.2 Dents

A dent is a depression in the pipeline caused by an external loading, producing a visible curvature/indentation to the wall of the pipe. A dent is considered smooth if it does not contain a stress concentrator (gouges, grooves, cracks, arc burns, etc.). Dents that exceed a depth of 6 mm in 101.6 mm OD pipe and smaller, or 6% of the outside diameter of pipe larger than 101.6 mm OD, should be considered defects. Pipe containing such defects should be repaired in accordance with ASME B31.8S or LI 2189 (latest edition), dents [8, 24].

In order to accommodate the smooth passage of internal inspection tools, dents that exceed the prescribed limits of the tools should be removed from the pipeline.

If a smooth dent is located on a mill or field weld and exceeds 6 mm in depth, it is considered a defect and must be repaired in accordance with ASME B31.8S or LI 2189 (latest edition), Dents. Replace all dents that affect the curvature of the pipe at the seam or at any girth weld.

5.3.3.2.1 Dents containing stress concentrators

All dents, which contain a stress concentrator, should be considered a defect. The stress concentrator should be removed by grinding until the ground area blends smoothly into the adjacent pipe. Pipes containing such defects should be repaired per ASME B31.8S or LI 2189 [8, 24]. If the defect is repaired by grinding, magnetic particle and ultrasonic inspections must be completed to verify that the crack has been completely removed.

5.3.3.3 Buckles

A buckle is defined as a localized distortion of the pipe wall, normally resulting from point loading on the pipe in a location of high bending stresses, and is characterized by creasing of the pipe wall. There is a deflection in the pipe axis at the point of buckling.

Buckling to any degree is considered a defect and must be repaired [8, 24].

5.3.3.4 Cracks (non-leaking)

It is a procedure that all pipe body surface cracks are defects. Pipes containing such defects should be repaired per ASME B31.8S or LI 2189 (latest edition) [8, 24].

5.3.3.4.1 Stress corrosion cracking

It is imperative to take immediate action upon detecting any signs of stress corrosion cracking. The pipeline system operator must promptly verify and assess the condition. Upon identifying any defect that requires repair or removal, it is recommended to take swift action to remedy the issue through permanent or temporary repair or removal. It is highly discouraged to proceed with pipeline operations until the problem has been fully resolved [3, 29].

5.3.3.5 Replacement pipe

The failed section of the pipe should be replaced with a section similar or equivalent to that of the existing pipe. The minimum length of the replacement pipe should be 2 m. The wall thickness of the replacement section should be equal to or greater than that of the failed section.

The difference in wall thickness between the existing and replacement pipe should not exceed 1.6 mm. The requirements of ASME B31.8S and LI 2189 should also be met. Current applicable requirements can be found in Clauses 12.15 and 12.16 of LI 2189 [8, 24].

Testing requirements for pipeline repairs are as follows:
–   The replacement section should be pretested to the same conditions as for a new pipeline.

If a segment of a transmission line is repaired by cutting out the damaged portion of the pipe as a cylinder the replacement pipe should be tested to a pressure required for a new line installed in the same location, and the test may be carried out on the replacement pipe before it is installed [8, 24].

A repair that is carried out by welding should be inspected per requirement of an appropriate standard:
–   For pipelines licensed for, or containing any amount of, H_2S pressure tested to 1.4 times the MOP.

- For pipelines licensed as sweet and not containing any amount of H_2S pressure tested to 1.25 times the MOP.

Pressure testing should not be less than 1 h in duration. A copy of the test chart and associated documentation (including a signed affidavit verifying the pressure test) should be filled.

Codes such as LI 2189 dictate additional requirements for pressure testing new pipelines, but these do not apply to pretesting replacement pipes:
- Complete the pipeline pressure test report and forward it to the supervisor
- Visual inspection of the weld for pinholes, weld cap alignment, and excessive amount of weld material
- 100% radiography of all repair welds

5.3.3.6 Polyethylene (PE) pipelines

In the event of defects within a polyethylene (PE), it is permissible to make permanent repairs to a PE pipeline by performing a cutout of the defective portion of the pipeline as a whole cylinder and replacing the cutout portions with pipe or flanges that meet the design requirements required of the system. Heat fusion should be used for the joining process.

It is also permissible to perform temporary repairs to a PE system using fully encircling clamps as approved by the manufacturer. However, temporary repairs must be replaced by a permanent repair within 1 year of the initial repair being made [3].

5.3.4 In-line inspection

As per the established prioritized schedule, the pipeline system operator is mandated to take prompt action in response to any detected damage or condition based on the results of a risk assessment and the severity of ILI indications. The operator must review immediate response indications immediately, while other indications should be reviewed on time. A response plan, which includes the methods and timing of the response, must be developed. For scheduled or monitored responses, the operator may choose to reinspect instead of examining and evaluating within the specified time frame, as per their discretion [8, 21, 24, 25].

5.4 Job planning and approval

The supervisor or a delegate must be involved in all aspects of job planning to ensure the work is carried out in a cost-effective, efficient manner.

Job planning must be done for all tasks whether they are large or small and should cover all aspects of the work to be done ranging from prejob meetings to final cleanup.

Approval should be obtained before ordering special equipment or specialized personnel.

The schedule must include lead time to order all necessary materials and labor required to complete the work.

A detailed plan must be submitted to the pipeline engineering group for consultation if the work involves any changes or work out of the ordinary realm of maintenance or repair work [3].

6 Integrity assessment

The integrity assessment of a pipeline is an extensive process that involves a thorough inspection of the pipeline, evaluation of the indications that arise from the inspections, direct examinations, careful evaluation of the results of the examinations, and finally determining the resulting integrity of the pipeline through detailed analysis.

6.1 Pipeline integrity assessment plan

To ensure the utmost integrity of your pipeline system, it is crucial to establish a comprehensive plan that covers the following vital items:
1. Develop a meticulous schedule for baseline inspection and continual assessment of all pipeline segments. Implementing a risk assessment according to Chapter 9 prioritizes the process to ensure optimal results.
2. Conduct thorough and precise inspections to identify possible pipe defects. To confirm the relevance of indications, direct examinations may be necessary.
3. Analyze and assess inspection results with precision to verify and demonstrate the fitness for the purpose of the pipeline.
4. Complete appropriate repairs on time as needed, without any delay.
5. Determine an appropriate reinspection interval, considering all factors and risks involved.

6.2 Pipeline inspection plan

It is imperative to develop a comprehensive inspection plan that encompasses a meticulous selection of inspection methods for each pipeline segment, inclusive of both baseline and continual assessment, defect assessment, repairs, and mitigation, as well as reassessment interval(s). This plan should be meticulously crafted to ensure that all aspects of the pipeline are adequately covered and to minimize the risks of defects and malfunctions that could lead to significant losses.

6.3 Choosing the appropriate assessment method

There are three main ways to evaluate the integrity, or condition, of a pipeline:
- Internal (in-line) inspections
- Direct assessment (DA) of the pipe
- Hydrostatic testing

https://doi.org/10.1515/9783111629742-006

To ensure the integrity of pipelines, it is crucial to select the appropriate assessment method for each threat category (see Table 6.1). The following methods have been recognized as effective integrity assessment techniques:
- pressure testing,
- in-line inspection (ILI),
- DA (which includes external corrosion DA (ECDA), ICDA, and SCC DA (SCCDA)), and
- visual inspection of aboveground pipeline sections.

For further information on these techniques, kindly refer to Sections 6.4–6.6.

When determining the appropriate assessment method, several factors must be considered. These include the possibility of taking the pipeline segment out of service for the expected duration of the inspection, which depends on the number of customers affected and whether supply disruption is an option. The pigability of the pipeline section is also important and determined by factors such as geometry, constraints related to the propulsion fluid (velocity control, availability and possibility of disposal of water, and quality of water), and the capability of aboveground pipe support to carry the weight of the water or other propulsion fluid. Additionally, the reinspection interval, presence of parallel pipeline (electrical interference), and transported gas (type of expected defects) are all selection criteria that must be considered [3].

Thorough and reliable integrity assessments are easily achievable when based on a comprehensive threat analysis.

Table 6.1: Pipeline integrity assessment methods.

Integrity threats	Integrity assessment method
Corrosion	
Internal corrosion	– In-line inspection – Pressure test – Operating and maintenance procedures
External underground corrosion	– Direct assessment – In-line inspection – Pressure test – Visual inspection
External aboveground corrosion	– In-line inspection – Pressure test – Visual inspection – Operating and maintenance procedures
Stress corrosion cracking	– Direct assessment – In-line inspection – Operating and maintenance procedures

Table 6.1 (continued)

Integrity threats	Integrity assessment method
Hydrogen embrittlement	– In-line inspection – Pressure test – Operating and maintenance procedures
Manufacturing defects	
Pipe	– In-line inspection – Pressure test* – Operating and maintenance procedures
Pipe seam	– In-line inspection – Pressure test* – Operating and maintenance procedures
Construction defects	
Girth weld	– In-line inspection – Pressure test* – Operating and maintenance procedures
Buckle/dent	– In-line inspection – Pressure test* – Operating and maintenance procedures
Wrinkle bend	– In-line inspection – Pressure test* – Operating and maintenance procedures
Equipment	
Control/relief valve	Operating and maintenance procedures
Incorrect operation	
Incorrect operation	Operating and maintenance procedures
Cold embrittlement	Operating and maintenance procedures
Weather and outside forces	
Lightning	– Direct assessment – Operating and maintenance procedures
Cold weather	Operating and maintenance procedures
Flooding/erosion	– Visual inspection – Operating and maintenance procedures
Subsidence	– Visual inspection – Operating and maintenance procedures
Landslide	– Visual inspection – Operating and maintenance procedures

Table 6.1 (continued)

Integrity threats	Integrity assessment method
Mechanical damage	
Third-party damage	– Direct assessment
	– In-line inspection
	– Pressure test
	– Operating and maintenance procedures

*Refers to the post-construction era.

Several factors, including pipe age, factory coating type, field coating type, factory and field coating condition, cathodic protection system compliance, casing, soil-air interfaces, and soil type, contribute to the susceptibility of pipelines to failure due to external corrosion (EC). Integrity assessments and P&M measures that address EC will offset these factors to a greater or lesser extent. Conversely, local effects such as stray currents and/or interference may exacerbate these factors [3].

The risk of internal corrosion (IC) in pipelines is very low. This is due to the dryness of the gas. If the dew point is less than or equal to 20 °F (−7 °C), the gas is considered dry.

By controlling factors such as specified minimum yield strength (SMYS), temperature, age of pipe, type of coating, and distance from the compressor station, the threat of SCC to pipelines is low [3].

Stress corrosion cracking (SCC) and embrittlement can occur internally in pipelines carrying carbon monoxide and synthesis gas.

Pipeline failure has been shown to be caused by certain historical manufacturing practices. The presence of seam defects, defects in the composition, hard spots, and factors that lead to manufacturing inconsistencies are of concern.

Certain historical methods of constructing and joining have been shown to contribute to pipeline failure. The presence of wrinkled bends, buried couplers, buried threaded joints, poor selection and application of weld coatings, and factors that cause suspect welds are of concern [3].

A review of historical equipment failure data shall be carried out to determine which of the following types of equipment have a history of failure on the pipelines:
– control valves (set point drift outside manufacturer's tolerances),
– pressure relief valves (set point drift outside manufacturer's tolerances),
– flange gaskets,
– O-rings, and
– gaskets and packing.

Damage to pipework caused by third parties can usually be divided into two distinct categories:

- normal operation, construction, vandalism, theft, sabotage, and/or terrorism can cause third-party damage, and
- immediate loss of containment or rupture.

Third-party damage can occur as a result of normal operation, construction, vandalism, theft, sabotage, and/or terrorism.

Incorrect operation of pipelines can result in operation outside the integrity operating window (i.e., fluctuating pressure, humidity, etc.). Causes of improper operation include

- inadequate operating procedures; and/or
- failure to follow pipeline operating procedures.

6.4 Assessment planning for pipeline integrity

The initial assessment to determine pipeline integrity is known as the baseline assessment. To ensure that all pipelines or pipeline segments are examined within the integrity management program, a baseline assessment plan must be created. This plan must outline when each line will be assessed and the method used for the assessment [3, 8, 21, 24, 25]. It is a critical component of the pipeline integrity management system. The baseline assessment plan should, at a minimum,

- identify the specific integrity assessment method(s) to be implemented;
- specify the schedule for conducting those assessments, and explain the technical rationale for selecting the assessment method(s) and the risk basis for establishing the assessment schedule;
- the schedule should be based on legal requirements, risk assessment results, past incidents, and specific localized features of the environment.

6.4.1 Continual assessment of pipeline integrity

Maintaining pipeline integrity is crucial for ensuring safe and efficient transportation of oil and gas products. The process of ongoing assessment and evaluation is an effective way to identify any integrity issues, and make necessary repairs and replacements in a timely manner. After conducting an initial assessment of a pipeline, subsequent evaluations consider the data collected during the previous inspection, and using the results of risk assessments, a plan for periodic pipeline integrity assessments is developed [3, 8, 21, 24, 25]. This plan includes the same components as the baseline assessment and helps determine inspection priorities and intervals. By comparing results from subsequent inspections, a more precise evaluation of current pipeline conditions and their changes over time can be made. This continual pipeline integrity evaluation process enables us to maintain the highest level of safety and reliability for the pipeline system [3, 8, 21, 24, 25].

6.5 Defect assessment/evaluation

Assessing and addressing any defects in pipelines per local codes/regulations and organization requirements are crucial to ensure pipeline safety and integrity. Compliance with codes such as ASME B31G, API 579-1/ASME FFS-1, and BS 7910 is of utmost importance. Defects may arise from various sources, including corrosion, dents, or other mechanical damage and cracks. While some metal loss due to EC may be considered acceptable, any damage that compromises pipeline safety and integrity must be promptly and effectively addressed to prevent hazardous situations [3, 8, 21, 24, 25].

6.6 In-line inspections

ILI tools are used to identify, locate, and measure anomalies. An instrumented vehicle/tool called "pig" is propelled through the pipeline and provides highly detailed information about the interior and exterior of the pipeline. These tools can identify changes in the wall thickness of the pipeline. Any changes in wall thickness identified can be classified based on the calibration of the inspection tool on pipe sections with known features. The accuracy of these inspection tools is steadily improving. However, there are limitations in the measurement of feature size (length and depth), feature circumferential position within the pipeline, and feature location along the pipeline [3, 8, 21, 24, 25].

There are three types of anomalies looked for:
- Metal loss (or corrosion)
- Cracks
- Changes to the shape of the pipe

The pipeline must allow for the passage of the tool and have specific equipment in place to handle the equipment required.

Based on the test results each anomaly must be assessed to determine if it could cause damage. Address significant issues and make required repairs to the pipeline. Many types of defects can be assessed using the correct ILI tools. A magnetic flux leakage (MFL) tool is an example of an inspection tool. It uses magnetic technology to detect, discriminate, and size metal loss in liquid and gas pipeline [3, 8, 21, 24, 25].

ILI represents a valuable tool for assessing EC, IC, SCC, mechanical damage, and construction defects in pipelines. To select an appropriate inspection tool, it is worthwhile to consult authoritative sources such as ASME B31.8S, NACE Publication 35100, NACE SP0102, and NACE SP0113. However, before conducting an ILI, it is critical to evaluate certain pipeline features, such as reduced port valves, bends installed back-to-back, short radius bends, unbarred and back-to-back tees, and the availability of launching and receiving facilities. Multiple diameters and multiwall thickness pipelines may pose sticking issues, which need to be addressed as part of the inspection process.

When selecting an inspection tool, several factors must be taken into consideration. These include the inspection tool history, detection sensitivity, sizing accuracy, location accuracy, ability to inspect the full length and circumference of the pipeline, anomaly classification, ability to indicate the presence of multiple cause anomalies, and requirements for direct examination to determine the general reliability of ILI. By evaluating these factors, pipeline operators can ensure that they select a tool that provides accurate and reliable results.

6.6.1 Smart pig

Intelligent pigging is an inspection method whereby an inspection probe (also known as a smart pig see fig 6.1) is propelled through a pipeline. When the smart pig travels along the pipeline, it gathers information or data such as the pipeline's diameter, curvature, bends, temperature, anomalies on the pipe's internal wall, metal loss, general corrosion, erosion corrosion, pitting, weld anomalies, hydrogen-induced cracking, and the presence and location of corrosion [3].

The advantages of the smart pig over the traditional method of pipeline inspection are:

1. Smart pig can clean and inspect the pipeline while it travels through the pipeline.
2. It allows pipelines to be cleaned and inspected without having to stop the flow of product.
3. Smart pig saves companies both time and money because it provides cleaning and inspection services at the same time.
4. Some smart pigs are equipped with GPS that can help in mapping a pipeline. This helps maintenance personnel save time and money by showing the exact location of a defect, instead of having to excavate a large area to reach a specific location in the line.

Smart pigging is a nondestructive evaluation technique that uses the following methods of ILI technology:

a. Ultrasonic testing (UT)
b. MFL

Figure 6.1: A smart pig.

6.6.2 Pigging facilities

In order to carry out internal pipeline cleaning and inspection, conventional pipeline pigging facilities should be provided at the respective ends of each pipeline.

The barrels of the launcher and receiver should be of the inline type with safety interlock end closure doors, each barrel being of sufficient length to accommodate internal cleaning devices (PIGS). The pig traps should be designed in accordance with the relevant codes of practice for pressure vessels (e.g., BS 5500). Sphere/barred tees should be fitted into the pipeline to allow free passage of the pigs at all locations where the branch to line ratio exceeds 40%.

Double valve isolation should be provided on flow, isolation, and kicker/bypass lines; normally these should be ball valves on piggable sections and gate valves on non-piggable sections. The method of operation for these valves should depend on their size and the operation philosophy for the pipeline having due consideration to plant emergency shutdown requirements.

Pig launching and receiving is normally a manual field operation and detectors. Pig signalers are usually installed to indicate the free passage of the pigs within the system [3, 8, 21, 24, 25].

With reference to the pig launcher and receiver, the hinged end closure should be provided with positive pressure interlocks to prevent the end being opened while the barrel is under any pressure whatsoever. The main line bypass and kicker line valves are usually all interlocked with visual instrumentation; for example, pressure gauges, being provided, allow only predetermined valve operation sequences.

A full range of secondary branches of suitable size should be provided on the launcher/receiver to accommodate standard facilities, that is, vents, drains, purge, safety instrumentation, and wash down requirements. No branch connection should be less than 1-inch nominal bore and should, in all cases, be provided with at least one isolating valve.

Bend radii associated with pig launcher/receivers should be compatible to the use to which the line is to be put with particular regard to the use of intelligent pigs and the minimum radius imposed thereby. Launch and receipt facilities should be sited on a paved area, which includes either a sump of adequate size to retain any liquid spillage when the trap is opened or be kerbed to provide a bounded area sufficient to retain any liquid spillage (see Figure 2.6–2.7).

Consideration should be given to the need for platforms and walkways necessary for the operation of the facility particularly on larger diameter pipelines.

6.7 Direct assessment of the pipe

Direct assessment (DA) is a structured process for pipeline operators to assess the integrity of buried pipelines. Inspection results are validated through integrity digs where you excavate the pipeline to evaluate it [3]:
–   DA is one of the three acceptable methods for evaluating the integrity of a pipeline segment.
–   DA may be used either as a primary or a supplementary method, implemented in conjunction with one of the other primary assessment methods, that is, ILI or hydrostatic pressure testing.
–   DA is limited to evaluating the risks of three time-dependent threats to the integrity of a pipeline segment:
    –   EC
    –   IC
    –   SCC
–   When a pipeline segment is scheduled for a full integrity reassessment at an interval longer than 7 years, confirmatory DA (CDA) may be used during the seventh year following a baseline assessment to verify or "confirm" the integrity of a pipeline from IC and IC threats only [3, 8, 21, 24, 25].
–   If External corrosion direct assessment (ECDA) finds pipeline coating damage, the operator must integrate the data from ECDA with one-call notification information and right-of-way information to evaluate the segment for the threat of third-party damage.

DA may be used to evaluate the integrity of pipelines for the following reasons:
–   where ILI or hydrostatic pressure testing cannot be used,
–   to avoid impractical, costly retrofitting of a pipeline,
–   to avoid interrupting gas supply to a station fed by a single pipeline, and,
–   to provide an alternative where sources of water for hydrostatic pressure testing are scarce, and where water disposal may create problems.

DA may provide a more effective, equivalent alternative to ILI and hydrostatic pressure testing for evaluating a pipeline's integrity.

6.7.1 How is direct assessment carried out?

6.7.1.1 External corrosion direct assessment

ECDA is a highly effective and nonintrusive method for assessing EC and third-party damage in pipelines. It requires minimal modification to pipeline components and can be performed without interrupting product delivery. ECDA can also serve as a supplementary method for other threats. However, it is important to note that it should only be used as an integrity assessment method if it is deemed feasible for each pipeline under consideration [3, 8, 21, 24, 25]. To perform a successful ECDA, one must refer to relevant resources like ASME B31.8S, NACE SP0113, and NACE SP0502. Although some factors may hinder the use of ECDA as an assessment method, such as excessive pipeline depth, lack of accessibility for excavation, unsuitable coating for aboveground surveys (e.g. plastic backed tape coating), pipeline running entirely under pavement or asphalt, interference from stray current(s), and bare or poorly coated pipe, ECDA is still considered one of the most reliable and cost-effective methods for assessing pipeline integrity [3, 8, 21, 24, 25].

According to Clause 14.64 of LI2189, an ECDA is a four-step process that combines pre-assessment, indirect inspection, direct examination, and post-assessment to evaluate the threat of EC to the integrity of a pipeline. ECDA requires:

1. **Step 1: Pre-assessment** – to gather and integrate data to determine the feasibility of using ECDA for a segment, the identification of ECDA regions, and the identification of two indirect examination tools to be used on the ECDA region.
2. **Step 2: Indirect examination** – to evaluate the pipe segment and identify indications of potential EC, to classify the severity of those indications, and determine urgency for their excavation and direct examination.
3. **Step 3: Direct examination** – to examine the condition of the pipe and its environment, to determine actions to be taken should corrosion defects be found, and to identify and address root causes.
4. **Step 4: Post-assessment** – to determine a segment's remaining life, its reassessment interval, and the effectiveness of using ECDA as an assessment method.

6.7.1.2 Internal corrosion direct assessment

As per industry standards, IC is not considered a major threat to industrial gas pipelines. Hence, conducting an ICDA on these pipelines would not be a practical approach. However, if the pipeline system operator suspects any IC, it is highly recommended to conduct an ILI tool or pressure test to ensure the safety and reliability of the pipeline system [3, 8, 21, 24, 25].

ICDA is a process an operator uses to identify any area along a pipeline where fluid or other electrolyte introduced during normal operation or by an upset condition may reside. The process identifies the potential for IC caused by microorganisms, fluid with carbon dioxide, oxygen, hydrogen sulfide, or other contaminants present in the gas (see Clauses 14.73–14.85 of LI2189).

Placing coupons, UT sensors at sensitive locations to monitor corrosion, on long-term basis is often the best practice. Baseline wall thickness data and then periodic UT scan at selected locations can also add value to the IC data accumulated through monitoring process.

For ICDA, there are four-step processes, based on the principle that liquids collect at the bottom of a pipe when a "critical angle of inclination" is exceeded for a specific gas flow velocity:

1. **Step 1: Pre-assessment** – to gather and integrate data and information to determine whether ICDA is feasible for the segment, to support the use of a model to identify locations where liquids may accumulate, and to identify where liquids may enter the pipeline.
2. **Step 2: CDA region identification** – to apply a specific model to identify elevation conditions and other pipeline fittings where liquids may accumulate.
3. **Step 3: Direct examination** – to excavate and examine pipe locations identified by the process as most likely for IC and to evaluate the severity of defects and remediate as code requires.
4. **Step 4: Post-assessment evaluation and monitoring** – to evaluate the effectiveness of the ICDA process, to monitor segments where IC was identified, and to determine re-assessment intervals.

6.7.1.3 Stress corrosion cracking direct assessment

Most industrial gas pipelines are generally not subject to significant threats from external SCC. Nevertheless, certain conditions must be present for SCC to pose a threat. These conditions include the pipeline being over 10 years old, non-fusion bond epoxy coating, product temperature exceeding 100 °F (37 °C), operating at or greater than 60% of SMYS, and being less than 20 miles (32 km) downstream of a compressor station. If the pipeline does not satisfy all of these conditions, it is not susceptible to SCC [3, 8, 21, 24, 25, 29].

It is worth noting that while these conditions do not guarantee the presence of SCC, the absence of any one of these conditions significantly reduces the likelihood of SCC. Therefore, it is important to assess all the parameters before determining the susceptibility of a pipeline to SCC. In conclusion, the conditions mentioned above are crucial in understanding the susceptibility of industrial gas pipelines to external SCC, and their assessment can help mitigate the risk of SCC damage to pipelines.

SCCDA requires a plan that provides for:

1. **Data gathering and integration**: To determine whether the conditions for SCC are present, requiring an assessment for SCC; to prioritize pipeline segments for assessment; and to gather and evaluate data related to SCC at all operator excavation sites. When all of the following conditions for high pH SCC are present, an assessment method must be applied:
   - operating stress greater than 60% of SMYS;
   - operating temperature greater than 100 °F;
   - within 20 miles downstream of a compressor station;
   - age greater than 10 years; and
   - pipe coating other than fusion-bonded epoxy.
2. **Assessment method**: To evaluate segments for the presence of SCC; determine its severity and prevalence; repair, remove, or hydrostatically test the valve section; and determine any further mitigation requirements.

When the conditions for SCC are present in a segment, the segment must be assessed and remediated, as specified in Appendix A3 of ASME B31.8S, applying:
- the bell hole examination and evaluation method, or
- the hydrostatic pressure testing method for SCC.

Applying CDA requires a plan specifying that CDA can only be used on internal and EC threats:
1. For EC, the plan must comply with 49 CFR Part 192.925, however:
   - only one indirect examination tool may be used, and one high-risk indication examined in each ECDA region; and
   - all immediate indications must be excavated in each ECDA region.
2. For IC, an operator's plan must comply with section 192.927; however, only one high-risk location must be excavated in each ICDA region.
3. When applying either ICDA or ECDA, if defects are found requiring remediation before the next scheduled assessment, the operator is required to apply the formula in Sections 6.2 and 6.3 of the National Association of Corrosion Engineers (NACE) Recommended Practice 0502 to schedule the next assessment.

6.7.2 Reference standards

In December 2004, NACE adopted the Recommended Practice 0204 for SCCDA and is now in the process of adopting a proposed recommended practice for ICDA. These should provide operators with additional guidance for addressing threats [3, 8, 21, 24, 25, 34].
- For ECDA: NACE RP 0502-2002, ASME B31.8S, and LI 2189.
- For ICDA and for SCCDA: LI2189 and ASME B31.8S, Appendices A2, B2, and A3.
- For ICDA: NACE SP0206, 49 CFR 192.927, and LI2189.

6.8 Hydrostatic testing

Hydrostatic testing, or hydrotesting, is one of several commonly used methods to verify the integrity, or condition, of the pipeline. Essentially, a hydrotest is conducted to check for strength, leaks, and tightness of a system, and to assess the fitness-for-service of a pipeline based on fracture mechanics analysis. The pipeline is filled and pressurized with water and then examined for leaks or ruptures [24, 25].

Pressure test (hydrostatic test and pneumatic test) is conducted above the operating pressure, often

$$1.1 \times \text{design pressure} \tag{6.1}$$

or

$$1.5 \times \text{design pressure}$$

6.8.1 Calculation of test pressure and duration

The basic calculation of pressure limit is established by primary Barlow's formula as modified by the governing regulatory or construction specifications:

$$P = \left(\frac{2S_H t}{D}\right) \times E \times T \tag{6.2}$$

where P is the design pressure (MPa), S_H is the hoop stress (kPa), t is the wall thickness (m), E is the longitudinal joint factor determined per ASME B31.8 and paragraph 2.15 of LI 2189 [8], and T is the temperature de-rating factor determined as per ASME B31.8 and paragraph 2.16 of LI 2189.

This may be supplemented with various variables like minimum and maximum possible pressure, based on the maximum operating pressure (MOP), maximum allowable operating pressure (MAOP), type of test elevation, and pressure affected by it.

6.8.2 Types of pressure test

6.8.2.1 Type by test media
a. Hydrotest: Although water is the most commonly used liquid for hydro-testing, any liquid that suits the pipeline condition can be used to pressurize the pipeline. A hydrostatic test gives you a snapshot of a single point in time, and it does not provide the probability of future issues with the pipe.
b. Pneumatic test: For pneumatic tests, media like air, nitrogen, and other gases are used for pressurizing the lines for testing. However, air is the most common media in this group. The pneumatic test is often used as a "leak test" for pipelines.

The test pressure is limited between 100 psi and the calculated pressure that would produce hoop stress of 20% of SMYS of the material as illustrated in Table 6.2. When pipelines are operating at less than 100 psi, then the testing should be at a pressure below 100 psi.

Table 6.2: Pneumatic test – maximum hoop stress during the test [24].

Test medium	Class location – 2	Class location – 3	Class location – 4
	Percent of SMYS	Percent of SMYS	Percent of SMYS
Air	75	50	40
Gas	30	30	30

6.8.2.2 Type of test by pressure and objective

Conduct a pressure test according to the ASME B31.8 standard, ensuring compliance with the criteria detailed in note 841.322 (d.f.) as shown in Table 6.3. This process entails carefully measuring and recording pressure levels to guarantee safety and adherence to regulations. The pressure test is classified according to the pressure and objective of the testing into three types as follows:

a.   Spike test: This is conducted for a very short duration at significantly higher pressure to determine the structural integrity of the pipeline. The test ratio is generally greater than 1.25 times the design pressure ($1.25 \times P$). The duration of the test ranges from 5 min to about an hour maximum, the time that would take to stabilize the pressure and inspect the line under pressure. Short duration of test is used to prevent any enlargements of subcritical anomalies preexisting in the pipeline. If the test pressure is maintained for the duration of the test, the test is deemed passed and acceptable.

b.   Strength test: This is used to establish the operating limits of a pipeline. Typically, the test pressure ratio is set to 1.25 times the MAOP ($1.25 \times$ MAOP). The duration of the test is minimum 4 h or longer. This test determines that there is no rupture or leak in the pipe material. If the test pressure is maintained for the duration of the test, the test is deemed passed and acceptable.

c.   Leak test (tightness test): This is conducted at or below the operating pressure to determine that a pipe segment does not show evidence of leakage. Typically, the test pressure ratio is less than 1.25 times the MAOP. The duration of the test is minimum 2 h or longer. The duration should be long enough to inspect for any leak. If a pressure drop is noted, and all such pressure variations can be explained, the test is deemed to have passed. The test is directly related to the change in volume in the test section. This change in volume can be due to various causes, such as

–   injection or withdrawal of test fluid,

–   release of pressure through relief valve,

    – dissolution of air in the test medium, and
    – change in calculated volume, due to external heat affecting expansion or contraction of the pipe.

6.8.3 Considerations for test media – water

The following conditions should be considered:
– Source of water
– Storage
– Need for any corrosion inhibitors (PH neutralization)
– Microbiological corrosion prevention (use of biocides)
– Volume required
– Fill rate
– Injection point
– Disposal

6.8.4 Considerations for test media other than water

Sometimes liquid petroleum may be used for tests:
– When using these liquids, the Reid Vapor Pressure should be less than 7 psia.
– Safety precautions and measures are associated with the use of liquid petroleum as a test medium.

Table 6.3: Test pressure limits as per ASME B31.8 (see note 841.322 (d.f.)) [24].

Class location	Permitted test fluid	Minimum test pressure	Maximum test pressure	MAOP, the lesser of
Class 1 Division 1	Water	1.25 × MOP	None	Test pressure/1.25
Class 1 Division 2	Water/air/gas	1.1 × MOP	None (water), 1.1 × MOP (air and gas)	Test pressure/1.1 or design pressure
Class 2	Water/air	1.25 × MOP	None (water), 1.25 × design pressure	Test pressure/1.25 or design pressure
Classes 3 and 4	Water	1.40 × MOP	None or design pressure	Test pressure/1.4 or design pressure

7 Pipeline operation and maintenance

7.1 Pipeline commissioning

Commissioning of pipelines occurs at the initial working life of a pipeline and after maintenance of the pipeline, which requires evacuation of the product. This section outlines the procedures available and required to commission a gas pipeline [3, 8].

7.1.1 Pre-commissioning

Pre-commissioning encompasses the activities following construction before the entry of gas into the pipeline. This includes testing of the pipeline and removal of the test medium, normally water. Preparation for commissioning involves getting the pipeline to the correct conditions for receiving high-pressure dry gas [12, 21].

7.1.2 Pressure testing

Pressure testing of pipelines using water is prescribed in the design code and often required by national authorities (e.g., HSE for the UK) as proof that the pipeline is leak-tight and has sufficient strength to withstand the test pressure, which is always above the maximum operating pressure. The normal method is to hydrostatically test the pipeline, as water is easy to pressurize, has no major effect if a rupture occurs, and similarly depressurizes easily. BS 8010 requires a hydrotest, whereas ASME B 31.8 allows for air or gas tests in location class 1, division 2, and class 2 areas up to a maximum of 1.1 times the design pressure [24]. When using air or gas as the test medium, recognition of the much larger quantity of stored energy in the pipeline must be taken into consideration during the safety assessment.

Filling of the pipeline uses pigs to seal the water from the air in front of it. For long pipelines, more than one pig is used, with the second following shortly after the launch of the first. It is common for one of these pigs to be fitted with a metal plate that has petals pre-cut into it and of a diameter of 95% of the minimum internal diameter. This plate is designed to show that the pipeline has this minimum diameter throughout its length and that there are no buckles or large dents present. If the plate arrives damaged, further runs are required with a caliper pig to identify the extent of the dent and to identify the location.

When the pipeline is full of water and no dents are found, pressure testing can commence. Temperature monitoring of the pipeline, at least three locations along the pipeline, is recommended as the majority of any pressure variation can normally be explained by temperature variations if the information is available. A successful pres-

https://doi.org/10.1515/9783111629742-007

sure test will result in the pipeline being approved for use by most regulatory authorities.

The different design codes allow differing levels of time the pipeline is required to hold the test pressure. ASME B 31.8 states at least 2 h, whereas BS 8010 states a hold period of 24 h as a minimum [24].

Acceptance criteria is defined for BS 8010 as acceptable if no pressure variation takes place, which is not accounted for by variations in temperature and the accuracy and repeatability of the measuring equipment. ASME B 31.8 does not provide any guidance in acceptance criteria.

7.1.3 Dewatering

Removal of the test medium, in most cases water, sometimes containing corrosion and bacterial inhibitors, is required as part of pre-commissioning. The use of inhibitors for water needs to be established with regard to overall corrosion rates and the source of the test water. The pipeline can generally withstand the general corrosion caused by potable water or seawater without significant corrosion taking place. However, without biocides, bacteria in the water can rapidly develop corrosion areas leading to pitting and a general increase in pipeline roughness when test water is left in place for more than 2–3 weeks. Many biocides are degradable and have a very low toxicity level. Information from manufacturers would need to be sought on this point before disposal can be agreed with the relevant authorities. The volumes and flowrates required for a large pipeline generally require the water is extracted and disposed of without the need for tankage or treatment before disposal. For example, a 100-km-long section of 36″ pipeline requires 65,000 m^3 of water at a flow rate of approximately 2,000 m^3/h, to achieve a 1 m/s flow required for the pigs to be effective [3, 8, 21].

Swabbing pigs designed to seal tightly against the side of the pipeline are sent along the pipeline, propelled by dried, cooled, and oil-free compressed air. The pigs will evacuate the majority of the water but will leave significant traces of water present. Foam pigs that can soak up the water are used in conjunction with swabbing pigs to remove the excess free water.

Alternatively, based on the amount of contamination acceptable, liquid pipelines often dewater the pipeline by pumping through with the product to be used by the pipeline and sending it to a tank to allow for separation of the water. Where this would affect the quality of the product, for example, for aviation products or where the end product is not available for use, dewatering as described above would be used.

7.1.4 Drying

For most gas pipelines and other pipelines that will be left empty awaiting final commissioning, drying of the pipeline is required. Drying of the pipeline, as opposed to dewatering, involves removing the remaining films of water and reducing the relative humidity of the air left to a point whereby the dewpoint approaches the minimum specification for the gas to be transported. The various methods available are described below [3, 8, 21].

The throughput of dry air will eventually reduce the relative humidity, but is not the most efficient means and requires large volumes of air being pumped through the pipeline.

Methanol or other hygroscopic liquids can be batched through the pipeline using swabbing pigs. The methanol plug of up to 500 m in length will pick up the remaining water, and the film left will evaporate at normal temperatures and be carried out with the propelling air. Problems can exist in obtaining sufficient quantities of methanol and disposing of it after passing through the pipeline. Vacuum drying applies a near perfect vacuum to the pipeline, continuously removing the water vapor that appears as the pressure falls.

Vacuum drying is relatively easy to perform, requiring a vacuum pump at one end, although if there is any quantity of water present, this can take a considerable time to remove it.

7.1.5 Commissioning

When the pipeline has achieved the required degree of dryness, the commissioning procedure can commence. If there is any delay between this and commissioning, a line fill of dry nitrogen is often used to prevent corrosion and keep the pipeline dry.

To commission the pipeline, a supply of product at a controllable pressure and supply rate is required. The main consideration in filling the pipeline is to be able to control the filling process and monitoring the product/air interface present at the boundary. To determine where this is and to limit the interaction to a certain degree, pigs are used to separate the gases [3, 8, 21].

For gas pipelines, air is vented in a controlled manner at the reception point, and monitoring devices are used to measure the gas content of the vented air mixture. When the pig is received or the gas content has risen to 95%, venting is suspended and the pipeline connected to whatever process equipment exists at the reception point or shut off awaiting commissioning of the remainder of the plant.

For liquid pipelines, the air is vented to start with, and when the interface between the liquid and air approaches the end point, the pipeline is vented via a tank with large venting points open to receive the liquid in what is often a frothy mixture of liquid before complete liquid becomes present.

7.2 Utilities

Many utilities such as water, gas, power, and telephone are likely to be encountered along the pipeline route. Prior to trenching, these utilities should be marked from the information provided by the respective authorities and service companies. In all cases, the immediate area around these services should be excavated in accordance with the requirements of the appropriate body. Under normal circumstances, the pipeline should be installed below the utility. A minimum separation of 150 mm, between the underside of the utility and the upper side of the pipeline, should be maintained. Where necessary, a concrete slab should be placed between the utility and the pipeline to limit possible physical damage during future operations.

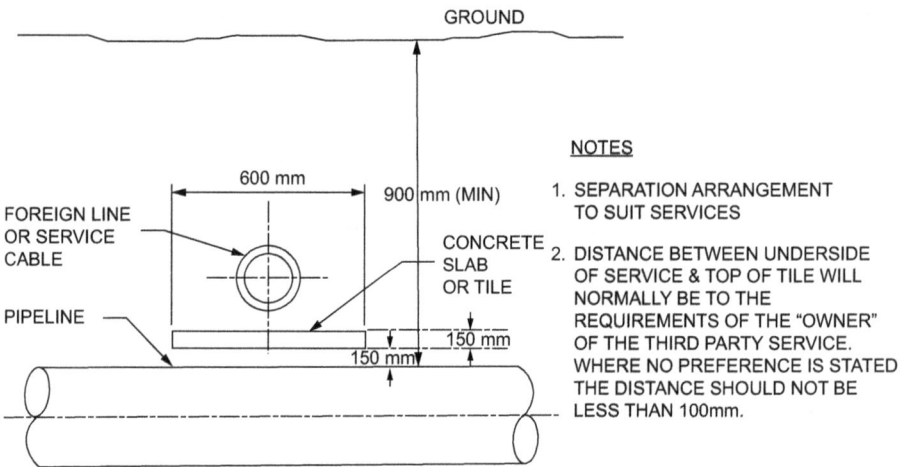

Figure 7.1: Typical third-party service crossing [3].

7.3 Control and monitoring

Sound engineering design, project-specific equipment, and comprehensive operating procedures should provide the basis for the control and monitoring of any petrochemical pipeline. All of these items should be addressed during the design phase, and when installed, they form an integral part of the commissioning and operating activities for the pipeline.

Depending on the operator's requirement either a pipeline integrity monitoring system (PIMS) or a supervisory control and data acquisition (SCADA) system may be designed and installed to provide constant monitoring of the system. Additionally, full metering facilities may be required for fiscal purposes [3, 8, 21].

7.3.1 Leak detection

Leak detection for the pipeline may be achieved by the provision of equipment (PIMS) to undertake constant monitoring of pressure sensing devices located at each end of the pipeline and at the intermediate isolating valve stations.

The output from these monitoring devices should be displayed in the main control room, and thereby enabling the operators to identify abnormal or unexplained deviations in pressure and to shut in the pipeline section affected by actuating the intermediate/block isolating valves.

In addition to the above, and any PIMS/SCADA system, metering should be installed at each end of the pipeline. The signal from the meter at the receipt terminal should be transmitted to the main control room (usually the pumping end of the pipeline) for comparison with the outgoing meter signal. Any unexplained deviation from a predetermined threshold value should alert the operators as to a possible leak or malfunction [3, 8, 21, 24, 25].

7.3.2 Corrosion monitoring

Internal corrosion of pipelines is not normally a problem as treatment to remove the corrosive elements within the product is usually undertaken at the inlet station/process plant. Where corrosive elements are known to exist and treatment is only limited or not possible, steels manufactured to NACE MR0175-91 and other appropriate standards may be used [3].

Notwithstanding the above and as an additional safeguard, internal corrosion monitoring of the pipeline may be undertaken. This monitoring could take the form of corrosion coupons and/or electrical resistance probes installed and monitored at each end of the pipeline.

Additionally, facilities may often be provided for the running of intelligent pigs. During commissioning of the pipeline, construction should be given to the running of such a pig in order to carry out and record a "baseline" survey of the condition prevailing inside the pipe. The record of such a survey should be stored and ultimately be used as the "base reference" for any similar or future surveys.

External corrosion may be monitored by assessing the performance of the impressed current cathodic protection system.

7.3.3 Equipment

Procedures should be developed during the design phase to ensure that all equipment associated with the pipeline control and monitoring facilities are periodically inspected. Valve actuators, local controls, and equipment in general, during these periods of inspection, should be serviced per manufacturers' procedures and good engineering practice [3, 8, 21].

7.3.4 Pipeline markers

Route alignment involves placing marker posts, generally at the extent of the working width and along the centerline of the pipeline, which is often not the center of the working width. Posts will be placed at all relevant crossings and all intersection points where the pipeline changes the direction. Landowners or occupiers will be required to ensure that crops are planted in such a way as not to obscure these items. Markers placed close to public highways or access points that become overgrown with vegetation should have the vegetation removed by the pipeline operator [3, 22].

At yearly intervals, the pipeline operator should carry out routine maintenance of these markers. All paints used during this regular maintenance and any other painted surfaces, for example, fencing, should be of a type that is not harmful to livestock.

7.4 Hydrate control

Refer to Section 2.12.1.1 for information on hydrate control.

7.5 Maintenance

Procedures for routine and preventative maintenance to include the various types of inspection described in this section should be established for the pipeline system. These should include the previously described pigging and surveillance facilities.

All equipment associated with the pipeline, end terminals, and intermediate valve stations should be periodically inspected and operated at scheduled intervals to ensure their continued effective functioning. Valve actuators, controls, and equipment should, during these periods of inspection, be serviced per the manufacturer's instructions.

Communication between the terminals and valve stations should also be tested periodically for the quality of transmission and operability [1, 3].

Pipeline marker posts should be properly maintained to ensure their continued visibility at all seasons of the year. Markers that become overgrown with vegetation should have the vegetation removed by the pipeline operator.

An emergency repair procedure should be prepared and held by the operator. This procedure should cover all actions necessary in the event of an emergency repair being required to the pipeline including notice to the appropriate statutory authorities.

A full inventory of spares, together with emergency repair equipment, should be specified. This equipment should be held at the main control center or a specially defined service base.

7.5.1 Physical status

Routine surveillance of the complete pipeline should be undertaken throughout its operational life. Primarily, this is to ensure the continued integrity of the pipeline by the early detection of any third-party activities on, or in the vicinity of, the route. Such surveillance should also be undertaken by a combination of both line walking and helicopter patrols, the frequency of such inspections will be subject to discussion, and agreement will be subject to discussion and agreement with the operator and the Department of Energy [3].

7.5.2 Line walking

At regular intervals the complete length of the pipeline should be inspected on foot by a representative of the operator.

To assist in this inspection, the pipelines should be delineated by pipeline marker posts, and typical examples of which are shown below.

7.5.3 Helicopter patrols

At regular intervals, a helicopter inspection of the entire pipeline route should be undertaken and video records made.

To assist the helicopter patrol in locating the pipeline, "aerial marker posts" should be positioned at strategic locations. To enable these markers to be easily visible to the helicopter crew, they should be designed with a reflective colored (day glow) surface (see Figure 7.2–7.3).

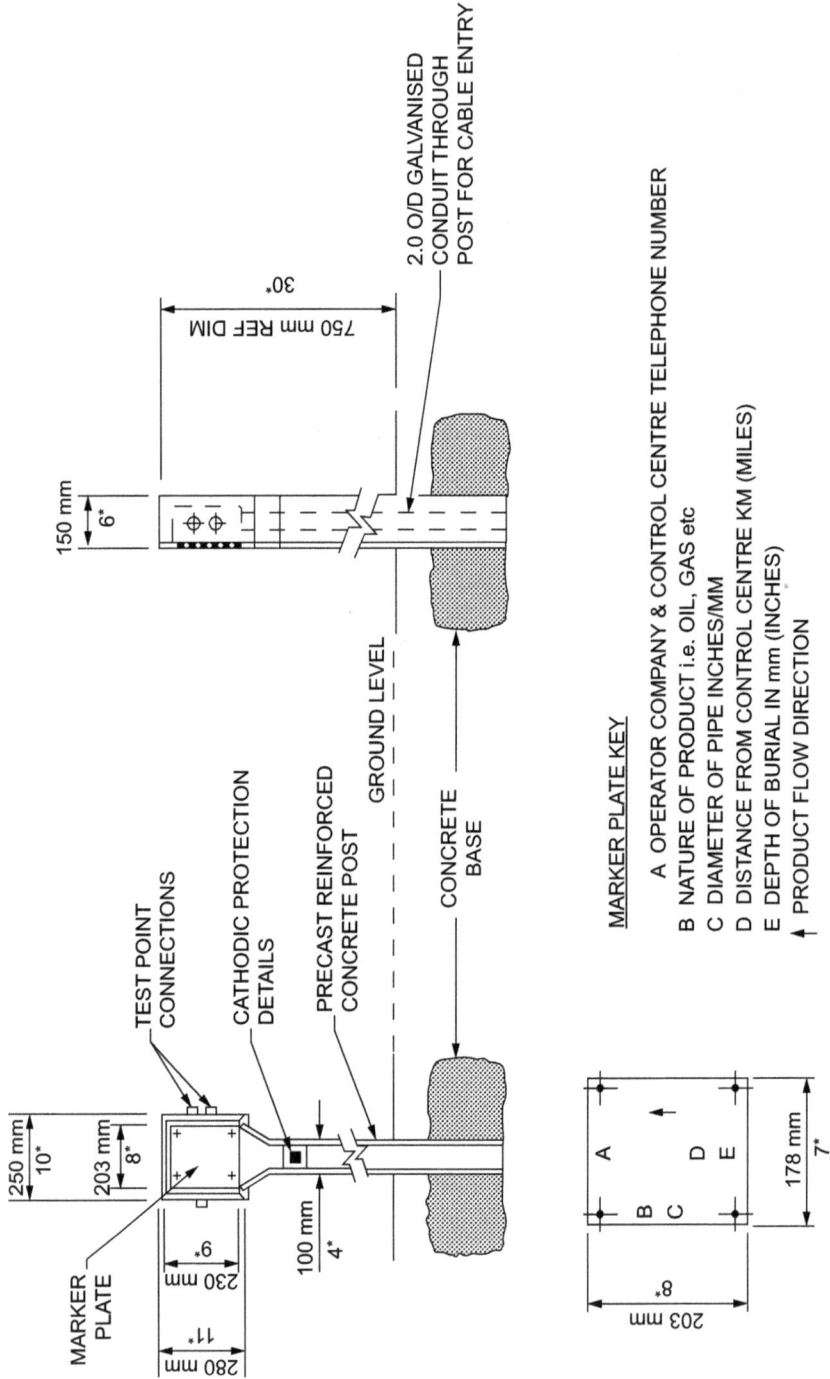

Figure 7.2: Pipeline marker and cathodic protection test points [3].

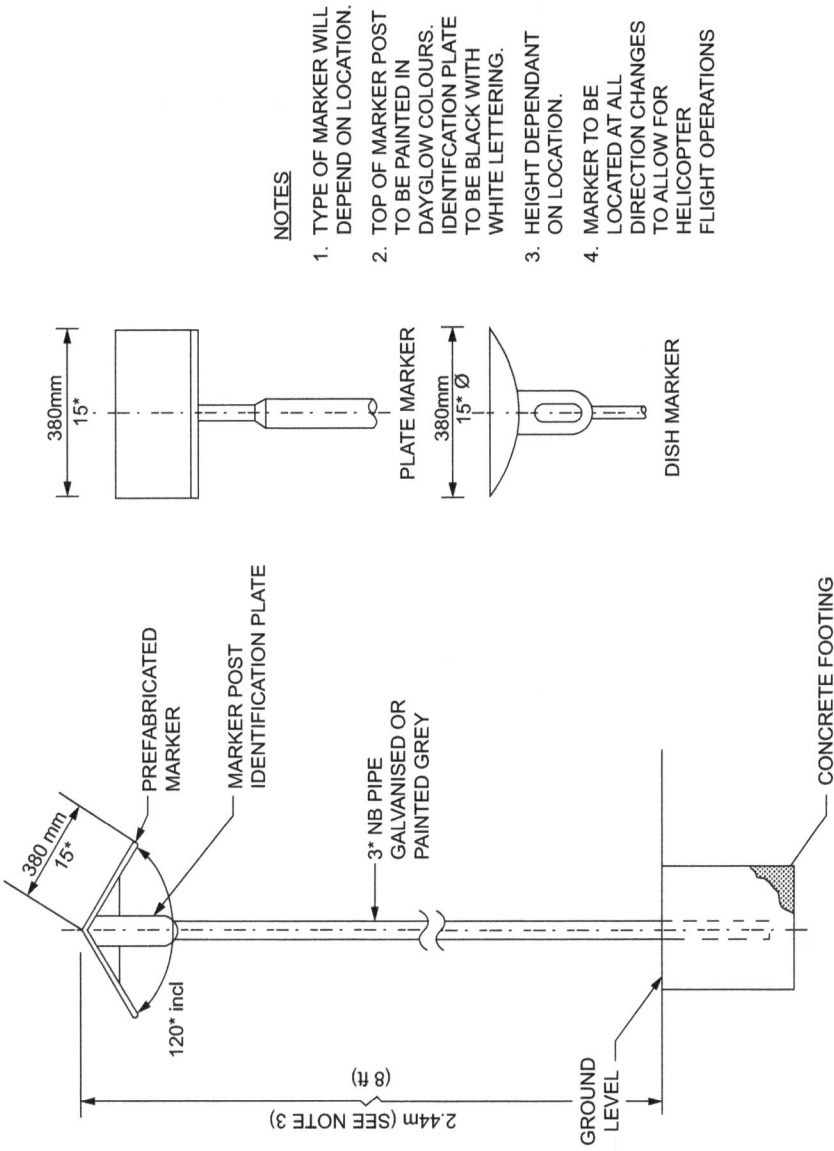

NOTES

1. TYPE OF MARKER WILL DEPEND ON LOCATION.

2. TOP OF MARKER POST TO BE PAINTED IN DAYGLOW COLOURS. IDENTIFCATION PLATE TO BE BLACK WITH WHITE LETTERING.

3. HEIGHT DEPENDANT ON LOCATION.

4. MARKER TO BE LOCATED AT ALL DIRECTION CHANGES TO ALLOW FOR HELICOPTER FLIGHT OPERATIONS

380mm
15*

PLATE MARKER

380mm
15* Ø

DISH MARKER

PREFABRICATED MARKER

MARKER POST IDENTIFICATION PLATE

380 mm
15*

120* incl

3* NB PIPE GALVANISED OR PAINTED GREY

2.44m (SEE NOTE 3)
(8 ft)

GROUND LEVEL

CONCRETE FOOTING

Figure 7.3: Typical aerial marker [3].

8 Records and data management

Operation and maintenance data are required for the documentation of Integrity Management Program activities as well as for risk assessment or technical assessment. Some data types, such as materials of construction and construction specifications, do not change over time, but permanent records are required to ensure that the correct data is referenced in all integrity activities. Other data sheets, such as regularly scheduled cathodic protection tests and chemical applications, require a historical database to track the data [1]. To create an accurate risk assessment of each pipeline system, collect data from the following types of sources:
- Paper records
- Electronic records
- Interviews with workers in the field
- Field observation and research

Data from the above sources can be compiled into a master table that serves as a reference for risk assessment and segment performance evaluation. To ensure the safety and reliability of pipeline systems, it is essential to have a well-documented integrity management plan. This plan must be updated regularly to reflect any changes in the pipeline, assessment methods, or areas of interest. Each update must be accompanied by a clear explanation of the reason for the change, an analysis of its consequences, and the approval of the relevant authority within the organization of the pipeline system operator. With an effective integrity management plan, pipeline systems can operate with confidence and minimize the risk of accidents or failures [1].

8.1 Types of data and records

8.1.1 Maps and drawings

Identification and location of pipelines and major facilities such as pumping stations, batteries, compression equipment, water junctions, highways, railways, main junctions, block valves, and rectifiers [1].

8.1.2 Technical data

1. Piping – Location and lengths for each pipe diameter installed, indicating wall thickness, grade and specification, backfill depth, working pressure and maximum working pressure, and pipe profile drawing.

https://doi.org/10.1515/9783111629742-008

2. Components – location, types, and pressure ratings of components. This includes critical block valves, ESDV or emergency shut-off valves, pressure switches, pressure safety valves, washers, and rectifiers.

8.1.3 Corrosion monitoring equipment

1. Intersections – Location and details of water, highway, rail, pipeline, and other intersections.
2. Special designs – Location and details of special designs and construction methods.
3. Pressure tests – Results of pressure tests on the new pipeline and any retesting results.

8.1.4 Licenses

All licenses associated with the gas pipeline should be kept on file with the relevant authority [1].

8.1.5 Repairs

Records of locations and details of leaks and required repairs should be maintained by the authority having jurisdiction over the design life of the pipeline.

8.1.6 Corrosion control programs

Current records that verify the operation and effectiveness of the corrosion control program for the piping system [1].

8.2 Data management

Pipeline data should be managed using a data management system such as a database. Pipe axis information and related data should be transferred from paper sources to the system.

Relevant pipeline data should be verified during the risk assessment process to ensure that the data is accurate and up-to-date for each pipeline in the system. Accurate and up-to-date pipeline data is essential for optimal risk assessment results [1].

The following pipeline attribute data refers to the physical characteristics of pipelines that do not change over time and should be kept in permanent records:
- Material specifications (e.g., pipe quality, size, wall thickness, coating type, and valve type)
  - Route and geography of pipelines (profile drawings and alignment sheet)
- Construction procedures (e.g., water transitions, paint installation, and welding procedures). Piping system performance data include but are not limited to [1]
  - Pressure, flow, and temperature data
  - Maintenance logs
  - Corrosion monitoring data
  - CP monitoring and messages
  - In-line control reports
  - Nondestructive testing and site evaluation
  - Personnel training and qualification records

8.3 Pipeline repair records

Performance data files contain cumulative data and score tables for a given relative data type. With each new set of data collected, data validation ensures compliance with existing criteria as well as comparison with historical data. Any nonconformance data will be flagged and prioritized for reassessment or corrective action. This activity may include data validation or operational and/or system changes based on new data findings. For example, if the last pig maintenance report shows abnormal pig wear, additional pig runs can be scheduled including a gauge check.

If the data is missing or of poor quality, one of the following actions should be taken: search and/or review other sources to locate the data, consult with a subject matter expert or other technical sources to determine a conservative estimate of the data, conduct fieldwork to determine, or confirm data or assume the existence of threats due to lack of data or poor-quality data. Keep records that identify how poor-quality data are used to assess their impact on risk assessment data. Maintain records that identify how poor-quality data is used to consider its impact on risk assessments [1].

9 Competency and training

This section outlines the procedures for maintaining the competence of personnel involved in pipeline integrity. Each area has the flexibility to combine various functions into a single position. In such cases, the position should possess the necessary skills as per the competency requirements outlined below. All suppliers are required to maintain records of their employees' qualifications per the integrity management manual of the organization. The responsibility for maintaining these training records lies with the contractor. Pipeline integrity engineers should have a good understanding of the relevant codes and regulations about pipeline operations. All pipeline integrity engineers must undergo integrity management training [1].

9.1 Requirements

- Insurance: Suppliers must furnish a copy of liability insurance.
- Worker's compensation: Contractors must provide a copy of their workers' compensation and injury assessment.
- Personal protective equipment (PPE): All personnel provided by the contractor are required to wear PPE, including approved steel-toed boots, hard hat, safety glasses, flame-retardant coveralls, gloves, and hearing protection. Additionally, employees must be clean-shaven to comply with respiratory protection requirements.
- Orientation: All contractor personnel must undergo HSE orientation before commencing work on pipelines or right-of-way equipment.

https://doi.org/10.1515/9783111629742-009

10 Management of change

The management of change (MOC) process is a critical element of any industrial management system. The introduction of any pipeline change, if not properly managed, can significantly increase the level of personal, environmental, safety, reliability, and process risks, thereby affecting pipeline integrity and potentially causing product delivery failure. To ensure that risks arising from any form of change are systematically identified, assessed, and managed, the MOC process should address technical, physical, procedural, and organizational changes to the system, whether permanent, temporary, or emergency [1].

Control of change procedures should be developed to identify and assess the impact of changes on piping systems and their integrity. The following points are essential to the proper implementation of the MOC process and integrity management program: all MOC procedures should be well understood by the personnel using them; the pipeline system operator should be aware that changes to the system may require changes to the integrity management program; and all affected workers should be involved in reviewing procedures with management as they can assess the impact on safety and suggest adjustments.

A documented record of changes must be created and maintained to ensure pipeline integrity. For any change, documentation should be included before and after the changes are made. All changes should, if possible, be stored in a central electronic system accessible to all relevant personnel. This information will help to better understand the system and possible threats to its integrity.

Communication of changes made to the piping system to all affected parties is essential for the safety of the system. Any changes to the system should be included in the information provided in the pipeline operator's communications to affected parties. In the event of changes, refresher training should be provided to ensure that personnel understand and adhere to current operating procedures.

Applications of new technologies in the integrity management program and the results of these applications should be documented and communicated to appropriate personnel and stakeholders.

Changes made to a pipeline system can affect multiple stakeholders within and outside the organization. Change management establishes procedures for identifying each and communicating with them when changes are made to the channel.

10.1 Objective

The objective is to ensure that changes that may affect pipeline integrity are identified, evaluated, and managed.

https://doi.org/10.1515/9783111629742-010

10.2 Requirements

Any physical change to the pipeline, whether permanent or temporary, and all proposed changes in maintenance, operational and management procedures, or organizational changes should be reviewed, approved, and managed through the MOC process. The impacts or risks associated with the change should be identified and addressed.

The affected personnel should be informed of the change and have the appropriate skills and knowledge to manage the associated hazards. Representatives of relevant disciplines should be involved to ensure that all aspects of the change are properly evaluated and that control or mitigation measures are put in place.

This includes identifying and incorporating changes in laws, regulations, and nonregulatory requirements affecting pipeline integrity [1].

Examples of changes in a pipeline system that require change management include changes that affect the:
–   Change of controlled documentation
–   Change in personnel or organizational structure
–   Pipeline ownership, change of operating licenses and permits, and legal and regulatory requirements
–   Pipeline safety devices and control systems
–   Operational condition of pipelines
–   Operating conditions that change
    –   Properties of production fluids (composition and volumes)
    –   Significant change in flow dynamics
–   Change in procedures, practices, standards, or policies
–   Change of location, route, etc.
–   Change in device configuration, size, type, drop configuration, etc.
–   Change in operational boundaries or conditions
–   Unwanted change or omission of safety devices
–   Change in process or design conditions

10.3 Temporary changes

It is of the utmost importance that temporary changes follow the same procedure as permanent changes with limited validity. Furthermore, after the specified time, the pipeline must return to its original state, unless a permanent change has been made [1].

10.4 Emergency changes

In scenarios where temporary or permanent changes cannot be made immediately, an emergency change will be initiated. This involves correcting any deficiencies that threaten the safety and health of personnel or the public, preventing immediate release into the environment, or resolving product supply disruptions. It should be noted that although emergency change may initially bypass the entire MOC process, it still requires careful consideration [1, 12].

11 Communication

To ensure that efforts to maintain integrity are effectively communicated, it is crucial to develop and implement a comprehensive communications strategy. This strategy should include regular updates and the provision of relevant information upon request. Our communication efforts should be frequent and widespread to ensure that all relevant parties and authorities are well-informed.

Our external communications should be strategically directed towards landowners and tenants along the right-of-way, public officials during non-emergency situations, local and regional emergency services, and the general public. We are fully committed to keeping all stakeholders informed about our integrity management efforts and their outcomes. Establishing and implementing a comprehensive communications strategy is essential to ensure that everyone is adequately updated on these initiatives. It is important that we provide consistent updates while also being ready to supply additional relevant information upon request. Our communication efforts should occur frequently enough to keep all involved individuals and authorities informed about the latest developments. We remain dedicated to ensuring that landowners, tenants, public officials, emergency services, and the general public are well-informed about our commitment to integrity management.

https://doi.org/10.1515/9783111629742-011

12 Auditing

12.1 Auditing and continual improvement

12.1.1 Evaluation process

The Integrity Management Program (IMP) should be periodically reviewed and audited to ensure that it meets its goal of improving safe and reliable pipeline operation. Once every year, your *Integrity Management* manual should be reviewed for content, adherence to regulations, and overall program effectiveness. In order to maintain a proactive approach to pipeline integrity, establish performance goals, which the IMP should meet or exceed. Conduct periodic field audits to evaluate the implementation and effectiveness of the IMP [1].

12.1.2 Performance measures

Performance measures should consist of internal corporate criteria as well as industry comparison criteria. The three types of performance measures are as follows:
1. Direct integrity measures – This includes leaks, ruptures, failures, and reportable incidents that result in a loss to operations or business.
2. Activity measures – This includes evaluating preventative or mitigative activities and determines the effectiveness of individual program elements. One example would be the total number of detailed water crossing inspections performed per year because of ROW surveillance inspections (see Figure 7.1).
3. Operational measures – This evaluates the direct impact of the implementation of the IMP on the pipeline system. This would include reducing internal corrosion rates at corrosion monitoring devices because of improved pigging and chemical treatment programs.

The minimum following performance measures should be reported to and reviewed on an annual basis:
- Number of leaks, failures, and reportable incidents (per unit length of pipe for industry comparison)
- Number of integrity assessments and length of pipe inspected
- Number of immediate repairs completed because of the integrity plan

https://doi.org/10.1515/9783111629742-012

12.1.3 Program audit

Annual audits should be conducted on the content, processes, and reference information contained in your integrity manual. The internal audit team should at a minimum comprise representatives of
- Operations department
- HSSE
- Integrity engineer/specialist

The audit scope should include but not limited to the following:
- Compare existing program practices to the current industry regulations and standards.
- Verify that personnel training and competency comply with the minimum requirements.
- Audit program records to ensure compliance with integrity management records and data management requirements.
- Compare most recent performance measures to corporate goals.

The audit should include reviewing at least two integrity projects and ensuring the personnel training and competency, data records keeping, and project documentation meet the minimum requirements.

The audit results should be documented and all deficiencies and corrective actions should be addressed via an action plan, which highlights the person responsible and action deadlines.

References

[1]   Anonymous, Pipeline Integrity Management Plan, Canada: Harvard Energy.
[2]   H. A. K. Hossam A. Gabba, "Framework of pipeline integrity management," *International Journal of Process Systems Engineering*, vol. 1, no. 3/4, p. 218, 2011.
[3]   M. S. Okyere, Mitigation of Gas Pipeline Integrity Problems, USA: CRC Press, 2020.
[4]   CAPP, Mitigation of External Corrosion on Buried Carbon Steel Pipeline Systems, Calgary – Alberta: Canadian Association of Petroleum Producers (CAPP), 2018.
[5]   K. A. Natarajan, Advances in Corrosion Engineering, Bangalore, 2012.
[6]   T. J. Langill, Corrosion Protection, Iowa: University of Iowa, 2006.
[7]   S. Kut, "Internal and External Coating of Pipelines," in *First International Conference on Internal and External Protection of Pipes*, University of Durham, 1975.
[8]   GSA, Natural Gas Pipeline Safety (Construction, Operation and Maintenance) Regulations – L. I. 2189, Ghana: Ghana Standards Authority (GSA), 2012, p. 85.
[9]   O. S. G. Singh, "Economics Criteria for Internal Coating of Pipelines," in *International Conference on the Internal and External Protection of Pipelines*, University of Durham, 1987.
[10]   CAPP, Mitigation of Internal Corrosion in Carbon Steel Gas Pipeline System, Calgary – Alberta: Canadian Association of Petroleum Producers (CAPP), 2018.
[11]   C. H. Klohn, "Pipeline flow test – evaluate cleaning and internal coating," *Gas*, August 1959.
[12]   B. March, Operations Integrity Management System, Imperial Oil, 2009.
[13]   R. M. Jorda, "Paraffin deposition and prevention in oil wells," *Journal of Petroleum Technology*, December 1966.
[14]   D. Zavenir, "zavenir-blog," 1 June 2018. [Online]. Available: http://www.zavenir.com/blog/types-of-corrosion-inhibitors/. [Accessed 22 September 2018].
[15]   Corrosionpedia, "Cathodic Protection and Anode Backfills," 15 September 2017. [Online]. Available: https://www.corrosionpedia.com/cathodic-protection-and-anode-backfills/2/1546. [Accessed 7 September 2018].
[16]   M. B. J. R. R. Fessier, "Pipeline corrosion. Final report," U.S. Department of Transportation Pipeline and Hazardous Materials Safety Administration Office of Pipeline Safety Integrity Management Program Under Delivery Order DTRS56-02-D-70036, USA, 2008.
[17]   P. M. Millar, "Asset Integrity Management Handbook," 2020.
[18]   E. M. Mark J. Kaiser, "Corrosion and Coatings," in *Pipeline Rule of Thumb Handbook (Ninth Edition)*, USA: Gulf Professional Publishing, 2023, pp. 285–351.
[19]   N. Fatmala, Offshore Pipeline Corrosion Prevention, Ocean Engineering ITB, 2016.
[20]   K. G. R. G. K. J. V. L. J. P. L. C. N. F. P. C. S. R. O. S. A. K. S. P. P. Michael Economides, "Production Engineering," in *Standard Handbook of Petroleum and Natural Gas Engineering*, Gulf Professional Publishing, 1996, pp. 363–983.
[21]   CSA, Z662-15, Oil and Gas Pipeline Systems, Canada: Canadian Standard association, 2015.
[22]   OECD, "Report of the OECD Workshop on Pipelines (Prevention of, Preparedness for, and Response to Releases of Hazardous Substances)," in *Organization for Economic Co-operation and Development (OECD)*, Paris, 1997.
[23]   Overpipe, "Third Party Interference (TPI) on buried networks," *International Pipe Line & Offshore Contractors Association*, 2018.
[24]   ASME, Gas Transmission and Distribution Piping Systems – B31.8, USA: ASME, 2010, p. 136.
[25]   IGEM, IGEM/TD/1 Edition 5 – Steel Pipelines for High Pressure Gas Transmission, London: Institute of Gas Engineers, 2008.
[26]   W. K. Muhlbauer, Pipeline Risk Management Manual: Ideas, Techniques, and Resources, Third ed., USA: Elsevier, 2004.
[27]   Weatherford, Line Pack Formulas, Weatherford.

https://doi.org/10.1515/9783111629742-013

[28] C. N. H. T. Reza Javaherdashti, Corrosion and Materials in the Oil and Gas Industries, Boca Raton, Florida: CRC Press, 2013.

[29] R. Singh, "Stress Corrosion Cracking in Steels," in *Pipeline Integrity Handbook: Risk Management and Evaluation*, USA: Gulf Professional Publishing, 2014, pp. 95–102.

[30] Corrosionpedia, "Cathodic protection monitoring," [Online]. Available: https://www.corrosionpedia.com/definition/6466/cathodic-protection-monitoring. [Accessed 17 January 2017].

[31] Q. B. Yong, Subsea Pipeline and Risers, London: Elsevier, 2005.

[32] H. Dean, "Special report-slick way to increase capacity," *Pipeline & Gas Journal*, pp. 17–19, June 1984.

[33] ISO15589-1, "Inspection and Monitoring," in *Petroleum and Natural Gas Industries – Cathodic Protection of Pipeline Transportation Systems – Part 1*, IHS, 2003, pp. 11–30.

[34] NACE, "Corrosion failures: Sinopec gas pipeline explosion," 2014. [Online]. Available: https://www.nace.org/CORROSION-FAILURES-Sinopec-Gas-Pipeline-Explosion.aspx. [Accessed 11 September 2018].

[35] S. Cato, "In-situ coating process using pigs, that can internally coat pipelines and carry out rehabilitation works," *World Pipelines: Coating & Corrosion*, pp. 7–11, 2019.

[36] Dairyland, "Dairyland – Solid State Decouplers," 2020. [Online]. Available: https://www.dairyland.com/knowledge-base-article/33-getting-started/106-what-are-solid-state-decouplers-and-how-are-they-used. [Accessed 17 January 2020].

[37] P. Lidiak, "Hazardous Liquids Pipeline Industry Perspective on Excavation Damage," in *Pipeline Safety Trust Conference*, New Orleans, 2010.

[38] Pipeline Safety Trust, "Pipeline Basics & Specifics About Natural Gas Pipelines," September 2015. [Online]. Available: http://pstrust.org/wp-content/uploads/2015/09/2015-PST-Briefing-Paper-02-NatGasBasics.pdf. [Accessed 28 August 2019].

[39] A. v. d. Werff, "The Importance of Pipeline Cleaning: Risks, Gains, Benefits, Peace of Mind," in *Pipeline Technology Conference*, Netherlands, 2006.

[40] S. Tong, Cathodic Protection, Sinopec, Ghana, 2015.

[41] M. Khalid, Tips for Effective Strategic Communication, AICPA & CIMA, 2020.

[42] J. B. P. K. Anna Turkiewicz, "The application of biocides in the oil and gas industry," *Oil & Gas Institute, Krakow*, pp. 103–109, 2013.

[43] A. Kadir, Distribution, Transmission, System and Design Lecture Notes, Salford: University of Salford, 2002.

[44] K. P. M. W. Tathagata Acharya, "Natural gas hydrates: A review of formation, and prevention/mitigation in subsea pipelines," *Advanced Science, Engineering and Medicine*, vol. 11, no. 6, pp. 453–464, 2019.

[45] O. Glaso, "Generalized pressure-volume-temperature correlations," *Journal of Petroleum Technology*, vol. 32, no. 05, pp. 785–795, 1980.

[46] D. Q. Kern, Process Heat Transfer, Echo Point Books and Media, 2019.

[47] D. W. Green, Chemical Engineering Handbook 8/E Section 1 Conv Factors & Math Symb (POD), 8 ed., McGraw Hill Professional, 2007.

Index

https://doi.org/10.1515/9783111629742-014

www.ingramcontent.com/pod-product-compliance
Lightning Source LLC
Chambersburg PA
CBHW081534220326
41598CB00036B/6432